Science and Fiction

D1319612

Science and Fiction—A Springer Series

This collection of entertaining and thought-provoking books will appeal equally to science buffs, scientists and science-fiction fans. It was born out of the recognition that scientific discovery and the creation of plausible fictional scenarios are often two sides of the same coin. Each relies on an understanding of the way the world works, coupled with the imaginative ability to invent new or alternative explanations—and even other worlds. Authored by practicing scientists as well as writers of hard science fiction, these books explore and exploit the borderlands between accepted science and its fictional counterpart. Uncovering mutual influences, promoting fruitful interaction, narrating and analyzing fictional scenarios, together they serve as a reaction vessel for inspired new ideas in science, technology, and beyond.

Whether fiction, fact, or forever undecidable: the Springer Series "Science and Fiction" intends to go where no one has gone before!

Its largely non-technical books take several different approaches. Journey with their authors as they

- Indulge in science speculation—describing intriguing, plausible yet unproven ideas;
- Exploit science fiction for educational purposes and as a means of promoting critical thinking;
- Explore the interplay of science and science fiction—throughout the history of the genre and looking ahead;
- Delve into related topics including, but not limited to: science as a creative process, the limits of science, interplay of literature and knowledge;
- Tell fictional short stories built around well-defined scientific ideas, with a supplement summarizing the science underlying the plot.

Readers can look forward to a broad range of topics, as intriguing as they are important. Here just a few by way of illustration:

- Time travel, superluminal travel, wormholes, teleportation
- Extraterrestrial intelligence and alien civilizations
- Artificial intelligence, planetary brains, the universe as a computer, simulated worlds
- Non-anthropocentric viewpoints
- Synthetic biology, genetic engineering, developing nanotechnologies
- Eco/infrastructure/meteorite-impact disaster scenarios
- Future scenarios, transhumanism, posthumanism, intelligence explosion
- Virtual worlds, cyberspace dramas
- Consciousness and mind manipulation

For further volumes:
http://www.springer.com/series/11657

Kevin R. Grazier   Stephen Cass

Hollyweird Science

From Quantum Quirks to the Multiverse

 Springer

Kevin R. Grazier
Sylmar, California
USA

Stephen Cass
Boston, Massachusetts
USA

ISSN 2197-1188   ISSN 2197-1196 (electronic)
Science and Fiction
ISBN 978-3-319-15071-0   ISBN 978-3-319-15072-7 (eBook)
DOI 10.1007/978-3-319-15072-7
Springer Cham Heidelberg New York Dordrecht London

Library of Congress Control Number: 2015930016

Cover image: © by Eric Chu, Paranoid Delusions, Inc. Vancouver, BC, Canada

Printed on acid-free paper

Springer is part of Springer Science+Business Media (www.springer.com)

KRG
To Dad. Couldn't have done it without your support.
SAC
To Annie

Foreword

Gentlemen, we can rebuild him. We have the technology…
Oscar Goldman, *The Six Million Dollar Man*

I was 6-years-old the first time I heard Oscar Goldman declare in the opening credits of *The Six Million Dollar Man* that Steve Austin would be the world's first bionic man. "Better than he was before. Better. Stronger. Faster."

And I believed him.

It was 1974, and the television landscape was dominated by landmark shows like *M*A*S*H**, *Little House on the Prairie*, *All in the Family*, and *Happy Days*. There was *Columbo* and *Kojak; Maude* and *The Odd Couple*. Fantastic programs all. But for 6-year-old me, the world stopped weekly at the countdown to the horrific crash that would forever change the life of astronaut Steve Austin, and spawn millions of kids running in slow-motion with the trademark bionic, "Nuh-nuh-nuh-nuh-nuh!"

With the exception of *The Planet of the Apes* and *Star Trek* re-runs, there wasn't much science fiction on television at the time, particularly Earth-based sci fi that was grounded in real science. Looking back, I don't think I even realized I was watching science fiction. The title sequence presented a near-perfect introduction to the concept of the show, complete with archival NASA launch video, high-tech medical graphics of cyborgian surgeries, and the dulcet tones of Oscar Goldman's voice-over assuring us, "We can rebuild him. We have the technology." Fiction or not, I bought into the concept hook, line and bionic sinker.

As the child of a scientist who had come *thisclose* to being selected by NASA to be one of the first scientist astronauts in space, I was the perfect audience. I spent summers messing with beakers in my dad's lab at UCLA; building model Estes rockets in our back yard; obsessively watching Stanley Kubrick's *2001*, and reading Ray Bradbury's *The Martian Chronicles*. I was certain by the dawn of the twenty-first Century we'd all have bionic parts and be flying rocket cars to work. Or Mars. Or both.

Boy, was I wrong.

Flash forward three decades to 2004. I did not have bionic legs. I did not have a flying car.

But I did have my own sci fi TV show.

It was called *Eureka*, and, like *The Six Million Dollar Man*, the concept of the series was grounded in theoretical scientific possibility. The show was set in the fictional town of Eureka, Oregon, a secret government think-tank formed by Albert Einstein and Harry Truman to house the greatest minds in science and technology. These eccentric geniuses were brought together to live and create technologies that were decades ahead of the rest of the world. Unfortunately, they would sometimes create things that threatened to destroy the world as often as save it. Enter everyman U.S. Marshal Jack Carter who stumbles onto the town during one such crisis, and his keen investigative instincts help him save the day. The reward for his effort is being reassigned by the Department of Defense to be the new town sheriff. Much to his chagrin.

In developing a show about a town of scientific geniuses, we quickly realized one very important thing: we were *not* scientific geniuses. This proved challenging. We knew the science had to be grounded if the audience was going to buy into the concept of the series (not unlike a certain show I had loved as a kid). We wanted the sci-fi element of the show to be a catalyst for the character drama, not the drama itself. The Sci Fi (now Syfy) Channel executives asked us to create a series bible covering the dos and don'ts of the show. This was our mission statement:

> We want to create a show that is first and foremost tethered by compelling, believable characters whose problems, passions and frailties mirror our own. These people just happen to have extraordinary talents which make their everyday dramas that much more interesting to watch… Our science, like our town, should always feel plausible. While we are mindful that "fiction" is a huge element of science fiction, the theories and devices created here need to be, on a fundamental level, based in real possibility.…

It was clear that achieving our goals would require the assistance of a *real live* genius. With our network's blessing (and, frankly, insistence), we hired a bona fide JPL rocket scientist to keep us honest. His task was to make sure that the science in our science fiction was as accurate as possible without compromising story. Did involving a scientist in the creative process limit our imaginations because of scientific limitations? No. Did we, on occasion, push the boundaries of scientific plausibility? Yes. But we made our best collective effort to be true to the science as well as the characters, and the stories we told were that much stronger for it.

This led to an invitation to do a panel at San Diego Comic Con hosted by *Discover* Magazine and The Science & Entertainment Exchange discussing, "Does Good Science Make Good Science Fiction?" Our moderator was science journalist Stephen Cass, and my co-panelist was *Eureka's* own intrepid science advisor, Kevin Grazier, the esteemed authors of this very book. The discussion was lively and the sentiment in the convention hall was clear: I was not a genius. But our audience accepted that *our characters were*, because we made the extra effort to get the science right. This allowed our viewers to suspend their disbelief and go along for the ride. With their loyal support and record numbers, the ride lasted for seventy-seven episodes.

Which brings us to the present, and to the uniquely entertaining book you now hold in your hands. Part analysis of the evolution of science fiction in popular media, part celebration of geek culture, part explication of basic science fiction principles, the authors explore myriad aspects of creating, writing and *appreciating* science fiction. It is a ridiculously ambitious endeavor that only Kevin and Stephen would voluntarily undertake. I only wish they had written it a decade ago when I really could have used it.

Maybe once we do finally crack time travel, I'll sneak a copy back to 2004 and leave it in the writers' room.

Kevin and Stephen, please get on that.

Or, more importantly, flying cars.

Jaime Paglia
Venice, California

Acknowledgements

We would like to thank all the people who have helped bring this book to fruition, starting with all the science fiction fans that provided the inspiration for this book. Whether online or in person at events such as the San Diego Comic Con or Dragon*Con, your questions and desire to engage with both the science and storytelling of TV shows and movies made this book happen.

Next, we are very grateful to all the science consultants, writers, and producers who gave so graciously of their time for interviews. By giving us your insights into the ground truth of how science fiction screenplays are written and then turned into actual productions, you have filled in a critical part of the story of how science is incorporated into science-themed fiction.

Our editor at Springer, Angela Lahee, deserves particular thanks for her early enthusiasm and later patience, especially when one book became two! A special thank you to Jaime Paglia and Michael Brotherton for their thoughtful foreword and afterword respectfully. Further thanks to Michael who, in his role as series editor, made many helpful suggestions that dramatically improved this work.

KRG would like to acknowledge the students from his "Science of Science Fiction" class at UCLA, who served as unwitting lab rats for much of the material in this book. Thank you to Jane Espenson whose uncannily-timed Twitter proclamations "Writing Sprint Commence!" has turned many a low productivity day into one I could be proud of. A special thank you to the writing team of Zack Stentz and Ashley Miller whose patience and generosity when it came to giving of their time for interviews was, by any measure, "Above and beyond."

SAC would like to thank his wife Annie Leist for her unflagging encouragement, perceptive comments on draft chapters, and accepting all sorts of odd media purchases under the rubric of "research"; Ada Brunstein for being an excellent Monday writing-sprint buddy; and his family in Ireland for all their support: Frank, Audrey, David and his parents, Stephen and Breda.

Kevin R. Grazier

Stephen Cass (on left)

Contents

1 Introduction 1

2 The Path to Nerdvana 7
2.1 I Can't Define Science Fiction, But I Know It When I See It 12
2.2 It's Alive! Fictional Science: History and Attitudes 13
2.3 Conceits, MacGuffins, and Tropes, oh my 25
2.4 Comparative Thresholds: The Point of Know Return 33
2.5 Hollywood Science and Science Education—The Condensed
     Matter Version 35

3 Hollywood Scientists: Reel and Imaginary 47
3.1 Scientist Representations 48
3.2 Scientists and Gender/Racial Demographics 58
3.3 Hollywood Scientists as Idiosyncratic Nerds 64
3.4 Hollywood Scientists as Active Villains 72
3.5 Modern Hollywood Scientist Archetypes 75
3.6 Is Asperger's the New Lab Coat? 80
3.7 Science as a Superpower 86
3.8 Scientist Representations II: Bigger, Badder, and Better Than Ever . . . 87

4 Matter Matters 91
4.1 Mass: The Sum of All Nucleons 92
4.2 Relativistic Mass 107
4.3 Does Antimatter Matter? 109
4.4 Giving Matter Mass: The Higgs Boson 110
4.5 Flame or Fusion? Chemical and Nuclear Reactions 110

5 Pure Energy 119
5.1 Feeling the Heat 128
5.2 Energy Generation: Synthetic Joules 131

6 Radiation: An All-Time Glow 139
6.1 Particulate Radiation 141
6.2 Electromagnetic Radiation 144
6.3 Ionizing vs. Nonionizing Radiation and Radiation Sickness 159
6.4 Why Is Radiation so Frightening? 162
6.5 So Where Did we Get the Notion of Giant Bugs and Reptiles? 163

7   **A Quantum of Weirdness**   169

   7.1   Catching a Wave   171
   7.2   The Age of Uncertainty   173
   7.3   Entangled: It Doesn't Happen to Just Rapunzel   178
   7.4   The Ultimate Computer   180
   7.5   Quantum Leaps   182
   7.6   Quantum Creations   188

8   **My God, It's Full of Stars: The Universe**   195

   8.1   An Even Briefer History of Time   200
   8.2   A Galaxy Not So Far Far Away: The Milky Way   208
   8.3   A Star-Studded Gala(xy)   210
   8.4   A Star is Born   212
   8.5   The Fault in Our Stars: Star Death   214

9   **Shortcuts Through Time and Space**   219

   9.1   Let's Do the Space Warp Again   229
   9.2   Jump To Another Realm   230
   9.3   Wormholes   232

10   **Moving in Stereo: Parallel Universes**   243

   10.1   The Simulated Universe   245
   10.2   The Multiple Big Bang or Quilted Model   247
   10.3   Braaaaaanes   248
   10.4   Hybrid Parallel Universes   254

11   **Braver Newer Worlds**   255

   11.1   Location, Location, Location: The Right Kind of Star   260
   11.2   Location, Location, Location II: The Right Kind of Galaxy   263
   11.3   Location, Location, Location III: The Right Kind of Planet   265
   11.4   If We Can't Find a Nice Place, Let's Build One   284
   11.5   Creating the World is the First Step, Now We have to Explore it . . .   289

12   **Afterword**   293

Appendices   295

Further Reading   307

1

Introduction

Man is unique not because he does science, and he is unique not because he does art, but because science and art equally are expressions of his marvelous plasticity of mind.
Jacob Bronowski, Scientist/Science Historian

Science arose from poetry ... when times change the two can meet again on a higher level as friends.
Johann Wolfgang von Goethe, Writer/Statesman

Science delivers "Wow!" moments like nothing else on screen. For viewers, science not only grounds a story in the natural world, it adds a special thrill that comes from the possible, the plausible, the believable, and the "what if?" While magic can be spectacular, no matter how hard we wish, there are no magic swords, magic wands, or cloaks of invisibility[1]. On the other hand, a crashed spacecraft looming out of the mists of an alien planet, a patient snatched from the jaws of death by a risky medical breakthrough, or a smug murderer betrayed by a few molecules left at a crime scene—why not? Screen magic does exist, and science can be one of its key spell casters.

Screenplays that rely heavily upon science must, therefore, be written with care, or the result may not be an "Oh, wow!" moment, but rather its arch nemesis, the "Oh, please!" moment. An "Oh, please" moment happens when events unfold on screen that are so at odds with how the real world works, or how the viewer perceives the world works, that they exceed the limits of artistic license. The viewer's willing suspension of disbelief is, itself, suspended. When an "Oh, please" moment occurs, viewers are forcibly expelled from the screenwriter's creative vision and emotionally cast adrift. No longer immersed in the narrative on screen, viewers become painfully aware that they are watching a cast of actors stomping around on sets where the paint is barely dry, and the craft services table just out of shot.

---

[1] Actually, the jury is still out on that last one.

years the accuracy of the science portrayed in productions for small screens has come under increasingly close scrutiny. There age of scientists, authors, science writers, bloggers, popular culture reporters, and even Monday Morning Quarterbacks, who have written about scientific accuracy (or, typically, its lack) in Hollywood productions. Sometimes the quarterbacking starts even before the final reel, as happened when astronomer Neil deGrasse Tyson criticized 2013's *Gravity* (Fig. 1.1) via Twitter. At science conferences, science fiction conventions[2], and science fiction conferences[3], presentations and panel discussions on the accuracy of science portrayed in Hollywood productions are ubiquitous. Some colleges and universities offer courses on "Hollywood Science"[4].

The National Academy of Sciences has even gotten into the science accuracy fray. Chartered by Congress in 1863 under an Act signed by Abraham Lincoln, the National Academy of Sciences' mission is to provide scientific advice to the nation. Located on the campus of UCLA, the National Academy's Science & Entertainment Exchange provides scientific consultation to the entertainment industry professionals with highly-trained scientists and engineers. The goal of the exchange is to "…use the vehicle of popular entertainment media to deliver sometimes subtle, but nevertheless powerful, messages about science"[5].

Still the issue of scientific accuracy in Hollywood productions is a very complex one, and not every scientific inaccuracy is an "Oh, please!" moment. Not only is the definition of what constitutes an "Oh please!" highly subjective, fictional shows and movies are not intended to be documentaries. Simply hiring a science consultant on the production is no guarantee of perfect scientific accuracy: one of the operational principles of the Science and Entertainment Exchange is that, in a conflict between story and science, story wins every time. The Exchange even counsels new science consultants that, "We are not the science police."

In fact, a wonderful irony is that while many have lamented the inaccuracies of the science in Hollywood science fiction, it is a genre defined, in part, by the inclusion of at least one key science inaccuracy—often many more in the case of TV and film science fiction. Nearly all science fiction stories rely upon at least one discovery or invention that *doesn't exist* in the annals of

---

[2] Author SAC moderated *Discover* Magazine's first "Science in Science Fiction" panel at San Diego Comic Con in 2007. A popular panel, it has been standing room only every year since.

[3] A science fiction conference, like the Eaton Conference held alternate years at the University of California, Riverside (http://eatonconference.ucr.edu/about.html) is an academic conference that discusses science fiction as a serious academic discipline. UCR is also home of the Eaton Collection (formally the J. Lloyd Eaton Collection of Science Fiction, Fantasy, Horror and Utopian Literature), the largest publicly-accessible collection of speculative fiction on planet Earth.

[4] Author KRG even created and taught one of these at UCLA.

[5] From their web site, http://www.scienceandentertainmentexchange.org/.

Fig. 1.1 Mission specialist Dr. Ryan Stone (*Sandra Bullock*) installs instrumentation to the Hubble Space Telescope in *Gravity* (*2013*). (Copyright © Warner Brothers Pictures. Image credit: MovieStillsDB.com)

today's knowledge. If this isn't the case, then the story at hand isn't science fiction, it's just a science-themed thriller, drama, comedy, et cetera.

So when scientists complain about how they and their fields of study have been portrayed by Hollywood, such complaints should be tempered with an understanding of story goals and constraints, as well as several different classes of common psychological biases that all people tend to share. Most scientists are not storytellers. Accomplished storytellers understand how to take a viewer into a fictional world and provide both plot and characters that viewers care enough about to watch for an hour or two—much longer in the case of serialized stories.

Conversely, good storytellers also know what will cause viewers to lose interest. It is often the case that screenwriters are being neither thoughtless nor stupid when they take shortcuts around fundamental truths depicting aspects of real world science—for example, the reality that scientific advances occur on the back of vast amounts of mind-numbingly detailed and tedious labor. Just as dramatic crime series rarely include scenes of detectives carefully filling out the mounds of paperwork essential to a successful prosecution[6], or arguing with their IT department over a broken printer, so, too, the practice of science must be compressed and condensed—especially on television where the writer for an hour-long drama has, in actuality, a mere 42 minutes to tell a complete story—one with a beginning, a middle, and an end—and which includes five act breaks for commercials. Just as the drama of human relationships is often heightened and accelerated, so too are scientific phenomena or the rate of technological processes often exaggerated. (If you can accept that, say, two ex-lovers can resolve major emotional conflicts over the space of a few hours or days while being shot at by international terrorists, then give a

---

[6] A notable exception being 2007's *Hot Fuzz*, which actually managed to make police paperwork funny.

enwriter who has a doctor administer a drug that works in
reen time, rather than in 20 hours.)

we will examine the science that appears in Hollywood pro-
oth the scientists' and the storytellers' points of view (we will
also make)ccasional foray into UK productions like *Doctor Who, The Hitch-Hiker's Guide to the Galaxy*, the under-appreciated *Space:1999*, and the vastly under-appreciated *Blake's 7*). One of us (KRG) has spent a career as a professional scientist and science consultant; the other (SAC) has spent a career observing scientists and engineers, and reporting on their advances to non-specialists.

It has been said that, "It takes a library to write a book." In addition to our research[7], we've also spent nearly 24 hours interviewing numerous writers, producers, directors, and top science advisors—the people who create the shows you love. We've also interviewed psychologists and pop media experts. One of the most enjoyable aspects of creating this book has been when our research has revealed that the biases we've brought into the project were erro-neous, and we've been led into other fascinating, and often counterintuitive, directions. One delightful surprise that we've encountered is that observa-tions we make as consumers of TV and film today, observations we think are unique to the high-tech VFX-laden productions of the twenty first cen-tury, often can be found in critiques of film from the early twentieth century. Sometimes everything new is old.

Our primary goal is not to excoriate the creators of movies and shows for errors[8], but to celebrate when they get it right. After all, despite mutterings from academics about how inaccuracies in shows and movies promote sci-entific illiteracy, more than a few scientists and engineers working today cite productions such as *Star Trek* or *Contact* as the initial inspiration—the incit-ing incident if you will—for their careers. These shows must have been doing *something* worthwhile. We are quite certain that, in a few years, university laboratories will be occupied by a cadre of students who were inspired to ex-plore the limits of the known, and push the boundaries of what's possible, by the likes of *Eureka* and *Fringe*.

We will also explore the nitty gritty that goes into balancing science and story in a screenplay, and will examine ways in which paying attention to science leads to better storytelling. We'll teach you the difference between a *trope* and a *MacGuffin,* why a screenwriter can have as many of the former as they like, but not the latter, and the many and varied ways in which scientific concepts and terminology are incorporated into scripts.

---

[7] Which, admittedly, meant (re)watching many of our favorite series, films, and DVDs which totally didn't suck.

[8] Though we will be pointing out some of Hollywood's most egregious scientific goofs early and often.

In doing so, this book is, admittedly, a piece of naked propaganda for having screenwriters and producers pay as much attention as they can to science. Science can be a tremendous well of inspiration, for both the big themes and the smaller twists and intricacies of a story. For example, the life cycle of real wasp species led to the chest-bursting antics of the eponymous extraterrestrial in 1979's *Alien*. The iconic scene in 1969's *2001: A Space Odyssey*—when astronaut David Bowman disconnects HAL 9000—owes a debt to an early demonstration of digital speech synthesis, even down to the choice of *Daisy Bell* as HAL's "closing number." The premises of several episodes in the reimagined *Battlestar Galactica* (2004–2009) were deeply informed by real astronomical phenomena, such as the dangerous levels of radiation that can be found in globular clusters (this threat was then used to motivate characters to perform desperate acts as they attempt to traverse a cluster). Medical and forensic procedural dramas regularly dip into real case studies to inspire and inform the plots of episodes.

Hopefully, this book may also help eradicate the one kind of inaccuracy that *is* guaranteed to drive us up the wall and leave us grumbling on our Twitter feeds and Facebook status lines. This kind of inaccuracy occurs when a writer simply repeats a myth that sounds like science, but isn't and, often, never was. Examples include: humans only use 10 % of their brains[9], people exposed to the vacuum of space will instantly explode or freeze, or that computers produce showers of sparks because of malicious or flawed software[10]. When these type of things show up, it's often not because a writer made a conscious judgment as a storyteller, but often because they were simply lazy—learning their "science" from previous works of science fiction, and propagating scientific myths in their own works.

Finally, we'll be spending a lot of time looking at how much screen science and technology overlaps with real science and technology, as well as how well Hollywood portrays scientists. By explaining some of the underlying scientific ideas and concepts that appear on screen, our wish is to give you, the viewer, the tools to recognize for yourself whether a production is depicting good science or bad. For scientists, we want to provide insight into understanding why Hollywood storytellers make some of the choices they do. After reading this book, you may find yourself being more critical of the science Hollywood serves you, resulting in a few more "Oh, please" moments, but we also hope you will gain a much deeper appreciation for the "wow" moments that remain that make them truly "Wow!"

---

[9] Known, literally, as the "Ten Percent Myth".

[10] Okay, true, once upon a time, when monitors were primarily made out of glass and metal—particle accelerators known to old folk as "CRTs"—it was possible to kill some displays by programming incorrect values directly into a video card, but the *computers* were fine.

2
The Path to Nerdvana

It is said that science fiction and fantasy are two different things. Science fiction is the improbable made possible, and fantasy is the impossible made probable.
Rod Serling, *The Twilight Zone (1962)*

The reality of nature is far more wondrous than anything we can imagine.
Neil DeGrasse Tyson, Host, *Cosmos: A SpaceTime Odyssey (2014)*

Science fiction is no more written for scientists than ghost stories are written for ghosts.
Brian W. Aldiss, Science fiction author and historian

In the Buddhist religion, *Nirvana* literally means "blown out," as one might blow out a candle or a match. Nirvana is a profound and imperturbable serene state of mind, arising when the fires of desire, delusion, and aversion are extinguished. It is a state devoid of suffering, and everybody's path to Nirvana is different. As we see it, in the Way of the Nerd, *Nerdvana* means "blown out" as in "mind blown." It is that profoundly excited, albeit transient, state of mind arising when the writer of a movie or television episode serves up a helping of badass—smothered in awesome sauce and seasoned with win— while extinguishing the fires of "Oh please!" and "Really?" It is a state devoid of suffering, and everybody's path to Nerdvana is different.

Both authors share a lifetime love of science-based, science-themed, and, in particular, science fiction movies and television[1]. Give us a bag of popcorn, a bottle of soda[2], a *good* space battle—one that avoids forays into the trite and pays homage to the laws of physics—and we have achieved Nerdvana. On the flip side, every time either of us encounters an egregious or easily preventable

---

[1] And books and magazines and comics and audio dramas and web series and, well, you get the picture. Our Nerd Fu is strong.
[2] Or Barry's Gold Blend tea for the Irishman.

science gaffe, a worn-out science fiction myth, or an overused trope, Nerd-vana becomes a far-off abstraction, and we transmogrify into Angry Nerds[3].

Sadly, most of the science portrayed in Hollywood TV and film over the years has been weird science, it just wasn't until 1985 that a filmmaker used the term explicitly. Rockne S. O'Bannon, creator of many well-known science fiction series (*Alien Nation*, *SeaQuest DSV*, *Farscape*) explains why: "In the past, the vast majority of the audience probably didn't have much of a foundation in what the specific science was in any particular show they're watching. It just had to pass the smell test. If it seemed real, if there was a sense of verisimilitude to it, then that was satisfactory"[4].

Shouldn't that be enough? Outside of documentaries, TV series and movies exist primarily to tell good stories[5] above all else, and storytelling has always included elements of the fantastic—ever since a *homo erectus* returned to the tribe without meat and told those gathered around the fire, "You should have seen the one that got away!" To sell the fantastic, storytellers learnt to use themes, events, and settings from the natural world to ground their stories in reality and imbue them with an air of plausibility, of *verisimilitude*.

The Epic of Gilgamesh, one of the most ancient works of literature[6], incorporates animals (both wild and tame), a forest, mountains, the weather, even a flood, basing the story in everyday reality while exploring themes of immortality, life, and death. With everyday elements like these faithfully described and woven into the fabric of the story—points of reference that most people have observed, or events to which they can relate—the story became an extension of events in the natural world, and so the audience was less likely to question *Gilgamesh*'s fire-breathing thunderbirds and scorpion men.

Is storytelling any different today? Not really. Despite the emergence of different forms of storytelling over the millennia—from Greek theatre to massive multiplayer video games—writers continue the tradition of grounding their stories in the natural world. A primary goal of the storyteller remains keeping the audience immersed in their story, and neophyte screenwriters are counseled to "never wake the audience from *your* dream." Anything that removes the audience from the writer's creative vision is to be avoided at all costs. Writer/producer Andre Bormanis (*Star Trek: Enterprise, Eleventh Hour*) explains, "The problem is that when you see something absurd, it pulls you

---

[3] You wouldn't like us when we're angry! It's a mixed metaphor, agreed, but true nevertheless.
[4] Grazier K (2013) The Science Advisor's Journey, in: Nelson D, Grazier K, Paglia J, Perkowitz, *Hollywood Chemistry*, pp 57–78.
[5] And generate revenue through tickets sales at the box office or advertisement time.
[6] Dating back to circa 1900 BC, perhaps even earlier.

out of the story. You are not in the world of the movie anymore. You're outside of it, commenting on it and being critical of it, because it's silly"[7].

Although the goal of immersion has always been a constant, the things that will pull an audience out change with time. The days of Shakespearean asides to the audience are over[8], or at least on long-term hiatus, but today there are many other less-obvious ways in which a screenwriter can pull the audience's attention out of the narrative. With ever more technically literate viewers, scientific gaffes in a film or television episode can transform each of them from being a willing participant, immersed in the writer's world in some distant future, to a person sitting in a room in the 21st century, arms crossed, saying "Oh, Please!" or "No way!" or "Really? Would that *really* happen? I don't think so!"

Northwestern University professor of biomedical engineering Dr. Malcolm MacIver, observes that the level of onscreen science fidelity that is acceptable to viewing audiences is, to some extent, culturally dependant: "Why is that realism is such a powerful force in movie making in North America and not so much in Europe? I think it has to do with, realism speaking to something in American thought that it doesn't, in Europe, [where] people are much more willing to have more poetic [story elements] and not have every little detail spelled out. That's something special, I think, that isn't characteristic of all people," says MacIver.

The answer might be found in science literacy. Although K-12 students in the United States consistently rank poorly when compared to students from other countries of the same age (and have been falling significantly behind other countries for at least the past 30 years), a 2007 study by Dr. Jon Miller, then of Michigan State University[9,10], found a silver lining: adult science literacy in the United States actually surpasses most of the rest of the world, including Japan and every European country except Sweden. Miller speculated that it is the stricter science requirements in U.S. undergraduate programs that are responsible for the difference. (Before celebrating, Americans should be apprised that science literacy in the United States is still very low—so celebrating this "victory" is akin to a snail winning a race amongst other snails and claiming, "I'm Usain Bolt!" There is still a very long way to go.).

---

[7] Grazier K (2013) The Science Advisor's Journey, *Hollywood Chemistry*, pp 57–78.

[8] Mostly over, at least in dramas, where breaking the fourth wall is much rarer than in comedy. One notable exception in dramatic science fiction is the *Doctor Who* 1984 episode "The Caves of Androzani." Trau Morgus, one of the episode's antagonists, performs several Shakespearean-like asides to the viewing audience. Longtime fans of the show still consider this to be one of the finest episodes ever. Outside science fiction there are, of course, the 1990 and 2013 versions of *House of Cards*.

[9] Ralof, Janet (2010) Science literacy: U.S. college courses really count, *Science News*, Vol 177, 6, p 13.

[10] Miller, Jon D. (2007) The Public Understanding of Science in Europe and the United States. Paper presented at the AAAS annual meeting in San Francisco (Feb. 16).

The science literate in the American audience seem to be helping to improve onscreen science accuracy, or at least the perception of accuracy, in Hollywood productions. Screenwriters today are under increasing pressure to play to these more literate viewers for two primary reasons. First, the public's understanding of science has come a long way since the birth of television. The degree of fidelity to science that was acceptable in previous decades is no longer acceptable today. Even the laziest viewer of the 21st century has a better idea what the surface of Mars actually looks like than the most dedicated scientists of 1963. The result is that dialogue that worked on the original 1978 version of *Battlestar Galactica* would have elicited an "Oh please!" if included in the 2003–2009 reimagined version. MacIver, who was the science advisor on the 2010 television series *Caprica*, and who also consulted on the films *Tron: Legacy* (2010), *Man of Steel* (2013), and others, adds, "People have a much greater thirst for science and technology [these days] and that is probably related to technology dominating our lives to an ever more increasing degree." The silver lining for producers here is that the very same viewers that might be highly critical of scientific inaccuracies in a screenplay can be vocal supporters of a show if they believe that the storyteller has acknowledged their intelligence and honored the science. There is an incentive to get the science right.

Second, the Internet has had a major impact. In days long past, when learned elders were the purveyors of tribal oral knowledge, inquiring listeners dared not overtly challenge the storyteller, and there was little or no recourse to do a fact check anonymously. Even in the latter half of the 20th century, a visit to the local library just to determine if an episode of *Star Trek* got some aspect of science right might often have been considered too arduous. On that score, the Internet has been a game-changer. Tom DeSanto, producer for both the *X-Men* and *Transformers* series of movies, says, "With the Internet, and the amount of research that we can do immediately … it used to be you had to go to the library and pull books. Now with this magic portal into the collective human consciousness, we can do a little more fact checking"[11].

Not only can the Internet provide an individual viewer with a flood of information, it facilitates a ripple effect in the audience as a whole. O'Bannon explains, "I think we're living in a very different world. The bar has been set higher. With the advent of the Internet, if people are interested/passionate in a show, they might be inclined to go online, seek out others who are also fans of the show, at which point they have access to the world. Others who are interested in the same show, may, in fact, have the advantage of some scientific

---

[11] Grazier K (2013) The Science Advisor's Journey, *Hollywood Chemistry*, pp 57–78.

knowledge and, therefore, [that knowledge] starts to encroach."[12] So even if a viewer fails to notice a science problem in an episode of their favorite show, it's sure that *somebody* else did and posted online about it—probably even before the end of the episode, thanks to social networks such as Twitter. (In a bid to boost first-run television viewing figures—i.e. to give people an incentive to watch their show when it first airs rather than streaming or download-ing it later—broadcast and cable networks have been fostering "social TV" technologies that make sharing opinions about a show in real-time as easy as possible, a double-edged sword if there ever was one.) If word spreads that the science in a television series or movie is "stupid," there can be a very real impact on that production's ratings or box office performance, longevity, and, ultimately, revenue.

Of course, science fiction isn't the only game in town when it comes to drawing upon science. We live in a technologically driven society. Conse-quently, scientific concepts and influences appear in nearly every genre seen on either television or the big screen. While it is true that there is no shortage of documentary TV series, movies, and even web content that (usually) por-tray science, scientists, and the culture of scientists correctly, or at least well, many of these are not produced in Hollywood. Biographical films, or biopics, usually, though not always, portray science accurately. On mainstream televi-sion, scientific themes, of wildly varying degrees of accuracy, have found play in procedural dramas like the *CSI* or *NCIS* franchises; medical dramas like *House, MD* (2004–2012); cop shows; daytime dramas; comedies[13]; and even cartoons. Still, science fiction is the primary torch-bearer when it comes to Hollywood's portrayal of science, and it's there that successes and failures in technical accuracy stand out in sharpest relief.

With the popular success of James Cameron's 2009 film *Avatar*, and critical acclaim for television's reimagined *Battlestar Galactica*—both of which were groundbreaking on several creative and technical fronts—the genre of science fiction has never been more popular or, arguably, more respected. With the possible exception of its close cousin fantasy, no other genre can compare to the variety of stories open to a science fiction writer. Heroes in works of science fiction have to go to greater lengths to emerge victorious. Average people, caught in extraordinary circumstances, have to overcome greater odds simply to survive. Lovers have to cross greater distances, in time as well as space, to be reunited[14]. Is it any wonder that so many of the highest-grossing

---

[12] Grazier K (2013) The Science Advisor's Journey, *Hollywood Chemistry*, pp 57–78.
[13] For Hollywood science (and science fiction) accuracy, it's surprisingly difficult to top *The Big Bang Theory*. They've got game.
[14] Amy and Rory Pond. Enough said.

films each year, and also of all time, are science fiction? The only thing that might be surprising is how long the genre has struggled for respect.

Another surprise is that it is television—for decades generally considered inferior to movies in terms of artistic credibility[15]—that has probably done most to improve science fiction's critical reception. Shows like *Babylon 5* (1993–1998), *Lost* (2004–2010), and *Battlestar Galactica* pioneered and established a highly serialized approach to television storytelling that has become the format of choice for many of today's best dramatic actors and writers.

2.1 I Can't Define Science Fiction, But I Know It When I See It

The historical struggle for respectability gave rise to various internal divisions within the science fiction community. Consequently, the term "sci-fi[16]" was used for decades in a mostly pejorative sense, distinguishing pulp science fiction publications and low-budget "B" movies from more serious works of science fiction. As the genre's critical reputation has improved, creators and fans have mercifully pretty much chilled out about this terminology and today "sci-fi" is, essentially, synonymous with "science fiction" [17].

What do we mean by when we say something is science fiction, though? As already noted, science and technology motifs permeate modern screenplays. Yet nobody confuses *House M.D.* for *Fringe* (2008–2013), even though both have been known to feature experimental medical procedures with highly uncertain outcomes performed by a mentally unstable drug user. When James Bond climbed into a space shuttle to infiltrate a super villain's space lair in 1979's *Moonraker*, did the movie cross over into science fiction? What about when Bond drove an underwater car in 1977's *The Spy Who Loved Me*?

When *The Encyclopedia of Science Fiction* was first published in 1979 it included a lengthy discussion and debate over the definition of science fiction, yet didn't converge on a lone conclusive definition. Currently, Wikipedia currently cites 30 separate examples under the entry "Definitions of science fiction."

---

[15] When Oscar-nominated movie actor John Lithgow decided to star in the sitcom *3rd Rock From the Sun* in 1996, pundits worried if he was committing career suicide. Now film stars like Matthew McConaughey, Claire Danes, and Steve Buscemi are as likely to be found on the small screen as the large.

[16] Science fiction Writer/Talent Agent/Superfan Forrest J. Ackerman first used the term sci-fi—similar to the then-popular term "hi-fi"—in the early 1950s. Although Robert Heinlien is sometimes credited with the first use of the term, he actually coined "sci-fic" a few years prior.

[17] In these pages, we will use "science fiction", "sci-fi", and "SF" interchangeably and as qualitative equals.

Taking the minimalist and tautological approach, controversial author Norman Spinrad offered, "Science fiction is anything published as science fiction." No help there, nor from author Damon Knight: "[Science Fiction] means what we point to when we say it." Legendary editor John W. Campbell suggested, "To be science fiction, not fantasy, an honest effort at prophetic extrapolation from the known must be made," but this definition suffers by defining the genre, in part, by what it is not. Science fiction legend Theodore Sturgeon claimed, "A science fiction story is a story built around human beings, with a human problem, and a human solution, which would not have happened at all without its scientific content." While the best yet, that definition could still be applied to today's procedural dramas like the *CSI* series, *NCIS* series, and medical dramas like *House, M.D.*—none of which are considered science fiction. Science fiction novelist David Brin quips simply, "Science fiction posits possibility that our children do not have to repeat our errors. That does not mean that all science fiction has to be optimistic." Literary scholar Tom Shippey believes, "Science fiction is hard to define because it is the literature of change and it changes while you are trying to define it".

So let us offer our own working definition for the purpose of this book: To be science fiction, (most of) the situations, natural phenomena, and technology in a story should be either plausible or based upon scientifically understood laws and theories at the time of the work's creation. In addition, the story must also make a departure into the unknown—there must be some element of *what-if*, however well-grounded. What if we could travel to another star? What if we could make intelligent machines? What if we could manipulate our own DNA?

2.2 It's Alive! Fictional Science: History and Attitudes

Both the French author Jules Verne and the British author Herbert George Wells have been referred to as "The Father of Science Fiction"[18,19], though Verne's writing career predates that of Wells by easily 40 years. Through both were pioneers of sci-fi, they had sharply different approaches, establishing a division which manifests in work produced to this day.

Verne was the more devoted to accuracy within his science romances: his first version of *Five Weeks in a Balloon* (1863) was rejected for being "too

---

[18] Adam Charles Roberts (2000) The History of Science Fiction, p 48 in *Science Fiction*, Routledge, ISBN 0-415-19204-8.
[19] The accepted term for science fiction stories in the era when these men wrote was "scientific romances."

scientific." Verne was also fascinated with exploration. In his day, explorers from many countries were eagerly surveying every unknown (to Europeans) expanse of the globe. In a twist on this largely horizontal activity about the Earth's surface, Verne often imagined vertical excursions, such as in *A Journey to the Center of the Earth* (1864), *From the Earth to the Moon* (1865), and *Twenty Thousand Leagues Under the Sea* (1870)[20].

H.G. Wells was less interested in rigorous adherence to science, and more interested in using fantastic stories as platforms for social commentary, a tradition continued by Hollywood with television series like the original *Star Trek* and the reimagined *Battlestar Galactica*. Take Wells' 1895 novel *The Time Machine*. Despite a veneer of scientific explanation (which Wells cribbed from an earlier short story he'd written about time travel), *The Time Machine* ducks exploring the implications of voyaging through time, such as the grandfather paradox or the manifestation of chaos known as the Butterfly Effect (both staples of later works of time travel). Instead, the novel is primarily a satire of the rigid class structure of 19th-century Britain, being mostly set in a far future when humanity has divided into the dim-witted but beautiful Eloi and the brutish but industrious Morlochs. It's basically *Downton Abbey* in the year 802,701 A.D.

> I prefer to see SF as a mirror to the present. Set up that mirror 50 years into the future and today's confusions become clearer.
>
> Brian W. Aldiss, Science Fiction Novelist

Dr. James Gunn[21], science fiction author, Professor of English, and founder of the Center for the Study of Science Fiction at the University of Kansas, shares his view:

> My thesis, at least, is that they [Verne and Wells] probably share the title of the "Father of Science Fiction," of their own kind. It seems to me that there are two major elements, or trends, in science fiction. One we can call the Vernian tendency of science fiction adventure and exploration and discovery, and there's the Wellsian version of philosophical consideration of change itself, and you can see those elements still existing in science fiction. There's a whole tradition of the Vernian adventure story, and the use of technology and exploration, and then there's the Wells idea story of philosophical considerations

---

[20] Verne also wrote a Faustian Romantic fantasy entitled *Master Zacharius or the Clockmaker Who Lost His Soul* (*Maître Zacharius ou l'horloger qui avait perdu son âme*), about a clockmaker whose intense pride leads to his damnation. In some respects, this was an 1850's Michael-Crichton-like cautionary tale about scientific hubris. The short story was adapted for television twice in 1961: for the *Shirley Temple Show*, and for *Alfred Hitchcock Presents*.
[21] Not the same James Gunn who co-wrote *Guardians of the Galaxy* (2014).

and notions and the exploration of concepts rather than the extrapolation of technology into the future.

Robert Sawyer believes that it is Wells who deserves the nod for overall title "Father of Science Fiction," explaining, "Wells was writing science fiction as social commentary … Verne was the Tom Clancy of his day, producing technothrillers, but without any cutting edge. Wells, though, wanted people to think: about the British class system when he wrote *The Time Machine*, about the evils of colonialism when he wrote *War of the Worlds* (1897). Science fiction isn't just escapism, thanks to him. If we'd only had Verne to build on, we might have had the original 1970s *Battlestar Galactica*, but we'd never have had the 21st-century reimagining".

Rod Serling, creator of *The Twilight Zone*, was cut from the same cloth as H.G. Wells, and wrote tales of science fiction and fantasy as vehicles for social commentary. He said, "On *The Twilight Zone*, I knew that I could get away with Martians saying things that Republicans and Democrats couldn't"[22].

> The original *Star Trek*, was never about the future, it was about the world of the 1960s. It was about the social issues that Gene Roddenberry wanted to address, but found that network people typically didn't like writers coming in with stories about God, or racism, or war. He thought, if I do it in a show that's about aliens and spaceships, they're going to think "Well, that's just fantasy, we don't really need to worry about what kind of stories he's telling on that show." *Star Trek* was very much a show about the present-day, but using the tools of science fiction storytelling to shine a different kind of light on those issues.
>
> Andre Bormanis, *Star Trek* Science Consultant

This fundamental difference in the writing philosophies between Wells and Verne is a recurring tension in science fiction, especially in Hollywood science fiction. Does a screenwriter present realistic science and technology, or focus on telling a compelling story with engaging characters, bending science to the needs of the story? The current thinking in Hollywood among writers and producers is that, while accurate science may be an important component to add verisimilitude, story trumps science every time.

Although Verne and Wells are among the most well-known early titans of sci-fi, and many Hollywood productions have spawned from their writings, they did not invent the genre. Both men appeared on the scene well after Mary Wollstonecraft Shelley, and in his book *Billion Year Spree: The True History of Science Fiction*, British author Brian Aldiss argues that Shelley's *Fran-*

---

[22] From the PBS series *American Masters*, episode "Rod Serling: Submitted for your Approval" (1995).

kenstein (1818) represents "the first seminal work to which the label SF can be logically attached." Despite having no formal education, Shelley was an avid reader, and the science depicted in *Frankenstein* is representative of the state of the art at the time the work was written, with reasonable extrapolations about what scientific advances might be possible. Shelley also wrote *The Last Man*, the first work in the "last man standing" type of stories—later examples include the novels *On the Beach* and *I Am Legend*, both of which were adapted into well-known movies, and the latter of which was the original inspiration for the modern zombie-apocalypse sub-genre[23].

However, the belief that *Frankenstein* represents the first work of science fiction is not universally held among SF historians. Gunn believes, "My own feeling is that *Frankenstein* is much too Gothic in its origins and nature." Some have argued that *The Epic of Gilgamesh* should be considered the first work of science fiction because it includes a graphic depiction of a flood scene, similar in feel to what one might find in modern disaster movies or apocalyptic science fiction. If we accepted this definition, then the movie *The Johnstown Flood* would be classified science fiction, and it clearly is not[24]. Further, though *Gilgamesh* includes text we could label as "science", if that was enough to classify it as science fiction, then all modern medical and forensic procedural dramas should be similarly classified as sci-fi as well. So science fiction without science is merely fiction, but fiction that includes some science is not necessarily science fiction.

> You can't have science fiction until you have the scientific method.
>
> Robert J. Sawyer, Science fiction novelist

Gilgamesh aside, there *is* a work that pre-dates Shelley's *Frankenstein* by nearly 200 years that has to be given serious consideration, one that both science fiction author Isaac Asimov and planetary scientist Carl Sagan have christened the first work of science fiction. The mathematician and astronomer Johannes Kepler, a contemporary of Galileo, wrote an edutainment story[25] entitled *Somnium* ("The Dream") between 1620 and 1630, published posthumously by Kepler's son Ludwig in 1634.

Somnium is the seventeenth century literary equivalent of a modern day docudrama—a discourse on physics and astronomy wrapped in a fictional

---

[23] *I Am Legend* itself has been adapted *four* times: *The Last Man on Earth* (1964), *The Omega Man* (1971), *I Am Legend* (2007), and the direct-to-video mockbuster *I Am Omega* (2007).

[24] There were actually three different movies about the Johnstown Flood, a 1926 animated short, a 1946 drama, and a 1989 documentary.

[25] Apparently Kepler was a pioneer in that genre as well.

narrative. In the story, Duracotus, a former student of Kepler's mentor Tycho Brahe, is transported to the Moon by the Daemon of Lavania. Over the span of the story, Kepler discusses concepts such as inertia, gravity, synchronous rotation and tidal locking[26], and Copernican astronomy. With the exception of the supernatural mode of transportation, the science described in *Somnium* was from the real cutting edge of what was known at the time (or reasonable extrapolations of it)[27].

Early filmmakers started shooting science fiction movies almost immediately after reliable film cameras and projectors were invented in the closing years of the 19th century. In 1892, French inventor Léon Bouly created one of the first motion picture viewing machines, which he called the cinematograph[28,29]. Without the capital to develop the invention, in 1895 Auguste and Louis Lumière[30] purchased Bouly's rights to the device, then applied the name to an invention of their own—a sort of combination camera and motion picture viewer. The Lumière brothers designed their cinematograph to improve upon what they viewed as flaws in Tomas Edison's early motion picture presentation machines. Also in the same year, early cinematic pioneer Robert W. Paul achieved the first successful projection of a moving picture onto a screen.

In May of 1895, H. G. Wells's novel *The Time Machine* was published, and Paul approached Wells about an idea that would make his work come to life. In October of that year, Wells and Paul applied for a patent that would have let audiences experience time travel as the protagonist did in the novel. The Time Machine would have been the first audiovisual multimedia production and, in principle, very much akin to the "4-D" experiences visitors can enjoy during Hollywood studio tours today. Baxter[31] writes, "As Paul and Wells conceived it, the Time Machine was the first audiovisual mixed media art form, a chamber filled with movable floors and walls, vents for injecting currents of air, and screens on which could be shown scenes from all periods of time by means of motion picture film and slides. Audiences could be given the illusion of moving back or forward in time, of seeing in close-up or at a distance life in eras long before or after their own. It was a sophisticated and showmanlike idea." As much as the terms "multimedia" and "transmedia" dominate the entertainment industry today, Wells and Paul had a jump on the concept by over a century.

---

[26] The phenomenon responsible for keeping the same side of the Moon always facing the Earth.

[27] Another early pre-*Frankenstein* work of science fiction was the 1752 short story *Micromégas* by French philosopher and satirist Voltaire.

[28] Greek for "writing by movement."

[29] Abel, Richard (2004) *Encyclopedia of Early Cinema*. Routledge, London.

[30] These are the same gentlemen who brought us *L'Arrivée d'un Train en Gare de la Ciotat* (literally, *The arrival of a train at La Ciotat Station*), featured in the 2011 film *Hugo* – a beautiful film, highly recommended.

[31] Baxter, John (1971) *Science Fiction in the Cinema*. Paperback Library, New York, p 15.

Baxter continues, "For the first time the imagination and mythopoetry of science fiction had encountered the pragmatism of technology. The relationship, though chronically strained, exists to this day." Though off to a promising start, the idea faded and died. Wells's eminence as a writer grew, and that consumed more of his time. Paul simply did not possess the capital to develop the invention. This had repercussions that influence cinematic science fiction today. Had Wells and Paul realized their original vision, cinematic science fiction would have had its genesis as a branch of the literary science-fiction, the two would share a common ancestor, and cinematic science fiction would likely be very different today[32].

Because The Time Machine never came to fruition, most science fiction and film historians cite Georges Méliès's 1902 short film *Le Voyage dans la Lune* (*A Trip to the Moon*) as the seminal example of the genre, but arguments can be made that two examples pre-date it, one also of Méliès's making. In the 1895 short *La Charcuterie mécanique* (The Mechanical Deli) by the Lumière brothers, men stuff a live pig into a large crate labeled "Charcuterie Mécanique / Craque á Marseille", and a robot—more akin to the assembly line robots in today's automobile manufacturing plants than an anthropomorphic automaton—converts the animal into various sausages and cuts of meat[33]. Two years later, the Georges Méliès film *Gugusse et l'Automate* (*Clown and the Automaton*) featured a circus clown creating an automaton. Upon its completion, the automaton grew in size, and slapstick ensued as the automaton chased the clown[34]. Clearly, neither film made any attempt to be cerebral.

Le Voyage dans la Lune was a sixteen-minute-long short film based loosely upon the works of both Verne and Wells[35]. It told the story of a group of astronomers[36] who travel to the Moon, explore its surface, meet its inhabitants, and return to Earth. These first science fiction films introduced automatons and space travel, story elements that have remained popular in science fiction to the present day (Fig. 2.1).

Although *Le Voyage dans la Lune* was wildly successful, amazing audiences with trick photography and special photographic effects, Méliès ultimately went bankrupt. His filmmaking career was beset by piracy, in particular of *Le Voyage dans la Lune* in the United States, by Thomas Edison and others (at the start of the 20th century, attitudes to international copyright in the

---

[32] More *Battlestar Galactica*s, less *Sharknado*s.
[33] Don't take our word for it, it's on YouTube.
[34] This film is not on YouTube, sad to say, and is presumed lost.
[35] It had identifiable elements from both, and also bears resemblance to Kepler's *Sominem*.
[36] Though Méliès had a penchant for science fiction filmmaking, perhaps because the genre allowed him to challenge himself creatively, he overtly satirized scientists, portraying his astronomers *Le Voyage dans la Lune* as inept, and assigning one the name of a seer (Nostradamus), one a magician (Alcofrisbas), and another an alchemist (Parafaragaramus).

Fig. 2.1 The iconic image from *Le Voyage dans la Lune* of the man in the Moon, struck by the craft that delivered the film's protagonists

Fig. 2.2 Production still from *Le Voyage dans la Lune*

United States were of a similar tenor to those prevailing in China at the start of the 21st). Méliès was also plagued by the criticism that the plots and characters in his films were eclipsed by special effects—so the next time someone complains about how movie directors care more about computer-generated imagery than a good script "these days," remember that the gripe is actually over a century old (Fig. 2.2).

It's important to understand at this point that although science fiction novels have been made into television series and movies[37], and movies and episodes of television series have been novelized[38], screen science fiction represents a "second genesis" of science fiction, one distinct from literary science fiction. While cross-pollination occurs, they are two different species (which can cause tensions between science fiction fans when screen science fiction is judged according to the standards and customs of literary science fiction, and vice versa). This, in turn, means the depiction of science and scientists varies greatly between written and broadcast media—very likely because literary science fiction, following in the tradition of Verne and Wells, is a product of the Industrial Revolution and the scientific enlightenment of the 19th century. On the other hand, the early science fiction filmmakers like Méliès and the Lumière brothers may have used the writings of Verne and Wells and others as source material, but their storytelling influences and motivations were quite different. On the origins of science fiction cinema, Baxter writes, "Its roots lie not in the visionary literature of the 19th century, to which science-fiction owes most of its origins, but in older forms and attitudes, the medieval fantasy world, the era of the masque, morality play, and the Grand Guignol"[39].

Not only did these early films frequently make light of science and scientists, German filmmaker Otto Rippert's six-part serial *Homunculus* (1916) was the first film in which both science and scientists are characterized as inherently dangerous—beyond that of a cautionary nature—a distinction which still divides written and cinematic sci-fi to this day. Science fiction historian Dr. James Gunn elaborates:

> Film science fiction is customarily traditional, in the sense that it is skeptical of scientific experimentation. Of course there are exceptions to that, but in general the impression you get if you see somebody experimenting in a laboratory in the opening scenes of a movie or TV program, you figure that something dangerous is going to emerge from this, and that the general philosophy of the science fiction film is that you should be very careful about what you experiment with, because you can unleash horrors upon your family, your neighborhood, your city, and even the world itself. Whereas written science fiction embodies more of the philosophy of science itself, in that it suggests that it isn't what we find out that threatens us, it's what we *don't* know and *don't* find out. So generally it encourages experimentation and study and finding out what's out there so we can use it, or defend ourselves against it, or prepare ourselves for it. In

---

[37] Often badly.

[38] Often badly.

[39] Le Théâtre du Grand Guignol ("The Theater of the Big Puppet") in Paris specialized in graphic naturalistic horror shows from the time it opened in 1897, until its closing 65 years later in 1962.

that sense, at least, when we compare the traditional science fiction film with the customary written science fiction, we have to understand, I think, that they don't embody the same kinds of principles.

There's another factor that affects how science themes play out in material intended for mass audiences. The unfamiliarity of new science or technology concepts tend to be compensated for by being packaged in very familiar story formats. In his 1970 book *Science Fiction in the Cinema*[40], John Baxter writes about how this has especially impacted science fiction on small screen, which has been historically dominated by the need of a relatively small number of broadcasters to appeal to the largest number of viewers, "The basic problem is one of caution on the part of producers. Its commercial possibilities in TV have always been uncertain. Producers, therefore, always hedge their bets when it comes to [science fiction]; no show is put out unless it contains, in addition to its [science fiction] elements, a selection gleaned from some other field".

For example, James Gunn's novel *The Immortal*, about a man who possessed immunity from every known disease, was turned into a series in 1969. After a brief exploration of the science fiction premise, it became a "car chase of the week" show, popular at the time. Gunn says:

> I think that, in general, Hollywood tends to do what it knows, and is good at. The fact is that there is a lot of money involved, and it tends to like prec-edent. [Producers are] interested in making a good film, and whether or not it is scientifically accurate, or science fictionally extrapolative, or exploratory, or thought-provoking, is far down the list, it seems to me, in their list of values. If they can do it, it's fine. If they can fake it, that's okay too.

Novelist Robert J. Sawyer, elaborates, "Once you get into television, it doesn't matter if you started with a scientific premise, [the science] gets lost in the shuffle, and it becomes one of the standard storytelling modes that Hollywood is comfortable with: the conspiracy show, the fugitive show, the procedural".

Thirty years after *The Immortal* was converted into a car chase show, Saw-yer's 1999 novel *FlashForward* experienced a similar transformation when it became the basis for the 2007 ABC television series of the same name. In the novel, a particle accelerator accident provides the genesis for an examination of the implications of the Copenhagen interpretation of quantum mechanics (on which we will elaborate in Chapter Seven, "A Quantum of Weirdness") When it was adapted for television by ABC it became a conspiracy theory

---

[40] *Baxter, John* (1971) *Science Fiction in the Cinema*, Paperback Library, New York.

show. Not coincidentally, conspiracy theory series like Fox's *24* were at the height of their popularity. Sawyer continues:

> Every conspiracy plot is interchangeable with every other conspiracy plot. You lose totally the sight of the fact that they're talking about a science issue. The conspiracy to assassinate the President of the United States in the first season of *24* [or] the conspiracy to replicate a particle physics experiment that might displace consciousness in the first season of *FlashForward*—it makes no difference what the backstory is, all the scenes are the same: this guy is kidnapped, that guy is kidnapped, somebody pulls a gun here, there's an explosion there, there's a chase there. You can't, on a beat-by-beat basis, tell those two stories apart—despite the fact that one is ostensibly science fiction, and the other is a counterterrorism show.

Before we collectively storm the offices of ABC, CBS, and NBC with pitchforks for such philistine attitudes, it's worth noting that the packaging approach adopted by network executives predates television. In 1871, George Tomkyns Chesney's novel *The Battle of Dorking*, a fictional account of a sneak invasion of England by Germany, was the first of a new genre of invasion novels. In the subsequent 43 years, until World War I, over 60 fictional stories featured the United Kingdom being attacked by one invader or another. The one we remember is H.G. Wells' 1898 novel *The War of the Worlds*[41], the first *alien* invasion story. Wells' embedded his science fiction critique of British colonialism within a story framework that was already popular at the time. There's a key difference though between *The War of the Worlds* and TV shows like *The Immortal* and *FlashForward* (so maybe keep that pitchfork handy): in *War of the Worlds*, the science fiction themes resonated throughout the novel, while they were given cursory examination, then mostly sidelined, in *The Immortal* and *FlashForward* (Fig. 2.3).

Although fans of science fiction are drawn to the genre, in part, because of the inclusion of real science concepts, Sawyer says that ABC network executives were very forthright in their belief that many television viewers are turned off if the science is too accurate or the show too technical, "The field that I'm known for, 'hard science fiction'—science fiction where the science is integral to the plot and is rigorous in its execution and extrapolation—is a non-starter on television. [On *FlashForward*] we were told every time we used a tech term, 25,000 people wouldn't come back after the commercial. Do that, you've got five commercial breaks in an hour, you've lost over 100,000 viewers … Peo-

---

[41] Wells wrote another invasion novel, *In the Air* (1908), which may have provided the inspiration for one of the first science fiction films, *Airship Destroyer* (1909).

Fig. 2.3 Joseph Fiennes and science fiction novelist Robert J. Sawyer on the Los Angeles set of *FlashForward*. Fiennes played FBI Special Agent Mark Benford, and Sawyer wrote the novel upon which the series was based. (Image Credit: Carolyn Clink)

ple, [ABC] thought, were turned off by the real science content"[42]. There is an old saying in Hollywood: "If you make people *think* they've thought, they will love you; if you make people *really* think, they will hate you." Unfortunately, there are enclaves where this sentiment still holds court.

Media expert Dr. Martin Kaplan, director of the Norman Lear Center at the USC Anneberg School of Communications, provides another reason why Hollywood executives take stories with science fiction premises and scrub the science from the fiction:

> I think Hollywood believes that it has a ring in its nose, and the chain attached to that ring is the audience. They will go wherever they think the audience wants them to go. And so if they think that there is reason to believe that the audience wants science fiction, or conspiracy, or car chases, that's what they're going to do. It's not because they like it, because they believe in it, that they have a commitment to it, it's only because that's the numbers they're getting. So I don't think Hollywood *wants* to do anything, Hollywood *is* the marketplace. It's an extension of the audience. All it wants to do is make money, and it has zero commitment to anything in the way of content by genre, by accuracy, by morality, you name it.

---

[42] Sawyer proudly added, "We had more science than any show on the air that year. Except *Big Bang Theory*, which totally whipped our ass."

Of course, repackaging science fiction stories into the genre of the day—whether it be superhero adventures or conspiracy plots—isn't the only way science and technology themes show up outside "pure" science fiction. For example, the particular appeal of medical dramas was understood since the early days of television's emergence into the mainstream: in his 1964 book *Understanding Media: The Extensions of Man*, media theorist Marshall McLuhan said that medical dramas would prove to be very successful because they "create an obsession with bodily welfare." Kath Lingenfelter, writer/producer for *House, M.D.* adds, "You're talking about the human body, which is something we all share, and something going wrong with it is a nightmare we can all share."

Considered to be the first medical drama series, *City Hospital* aired on CBS from 1951 through 1953. Although traditional medical dramas have remained popular ever since, shows like *Marcus Welby, M.D.* (1969–1976), *General Hospital* (1963–), and *ER* (1994–2009), have in more recent years had to share the screen with procedural medical dramas like *House M.D.*. This trend towards the procedural has been even more pronounced with cop shows, with forensic dramas like *CSI: Crime Scene Investigation* (2000–) and *NCIS* (2003–) colonizing a landscape once dominated by dramas in the style of *Hill Street Blues* (1981–1987) or *The Shield* (2002–2008).

In these procedurals, science and scientific methods are integral to the stories. The focus is less on the characters, and more on the series of events that depict how an ailment is diagnosed or how a crime is solved, generally employing high-tech equipment and techniques.

Despite not being "science" shows per se, procedural dramas are not immune to science accuracy issues and viewer complaints, complaints that often focus on the rate at which a process—like DNA sequencing, fingerprint matching, and facial recognition—happens. The so-called "CSI Effect," is a catch-all term for several different, sometimes even contradictory, influences of the popular *CSI* procedural drama on juries and crime victims[43].

Many of these crime and medicine shows make a point of showcasing their primary science consultants as part of their marketing efforts. It's also not unusual to see their producers, screenwriters, and even cast members boast of how they collaborate with science advisors to improve screen plays, something which also goes for many modern science fiction shows as well.

This willingness to share some of the limelight with advisors is in contrast to the culture found in movie production, where directors still often appear

---

[43] The CSI Effect is also credited with increasing the number of students enrolled in undergraduate forensic study programs. While some other manifestations of the CSI Effect are debatable, this one seems legit.

uncomfortable with presenting a film as anything other than their own singular vision, although science advisors are now sometimes included as part of promotional campaigns as a way of boosting a movie's perceived verisimilitude. (*Game of Thrones* mastermind George R.R. Martin has stated that he believes the director-centric paradigm that has dominated Hollywood movie production since the late 1970s, versus the writer-centric paradigm found in television, is a significant reason for why television appears to have eclipsed movies creatively.).

2.3 Conceits, MacGuffins, and Tropes, oh my

Scientific fidelity notwithstanding, viewers watch television and film because they want to escape, they want to be entertained, they want go along for the ride. Given the different manifestations of fictional science we've already explored, viewers also expect to allow for some artistic license, since there is also a reasonable expectation that the work is, in part, inaccurate or at least highly speculative. Kaplan, agrees, "People are passionate about entertainment, and they're passionate because entertainment is supposed to stir up your passion, it's supposed to make you a little crazy—you have to believe that made-up things are so real that you should care about them as much as you do." This willing suspension of disbelief[44] has its limits, however.

The fantastic elements of a story are known as *conceits*. A narrative conceit is a storytelling device—often one that's fanciful or far-fetched for the sake of interesting viewing—which helps drive the plot forward. It's essentially the core *what-if?* in our definition of science fiction above, the idea from which everything else flows. In Kepler's *Somnium*, the conceit was: if a demon could take us to the Moon, what would we see there? As a storytelling device, voyage by demon isn't that much different than travelling via the transporters of *Star Trek*.

Jon Amiel, who has the distinction of directing one movie (*The Core*, 2003) that is loathed by many scientists, and another that is just as equally loved by them (*Creation*, 2009), says, "You take something like *The Fantastic Voyage*. The idea of miniaturizing [the main characters] is what you call the 'gimme'. The conceit. The great stretch that you're asking the audience to take along with the storyteller. Once you've got past that "gimme", you can then proceed to follow your story with immaculate truth, if you choose. Sometimes you

---

[44] The term "willing suspension of disbelief" was originally coined by English poet Samuel Coleridge in 1817, no stranger to the fantastic himself.

Fig. 2.4 Director Jon Amiel on the set of *The Core* (2003). Copyright © Paramount Pictures. (Image credit: PhotoFest)

need to make that fictional leap in order to illustrate some bigger, broader truth" (See Hollywood Box: Story Conceits in Science Fiction).

The nature of story conceits also allows us to clarify the difference between a science-themed procedural drama and a science fiction procedural, i.e., between fictional science and science fiction. What separates *CSI* from *Fringe* or *NCIS* from *The X-Files*? While story conceits are alive and well in procedural dramas like all the *NCIS* and *CSI* series, what makes these shows non-science fiction is the nature of the conceits. In science fiction, the conceits are about new or different scientific ideas. In forensic procedurals, the science is taken as established: the conceit is rather in pretending that real-life limits on time and police budgets don't exist (Fig. 2.4).

Bradley Thompson, who, along with his writing partner David Weddle, wrote for *CSI* between stints on *Battlestar Galactica* and *Falling Skies*, explains that on *CSI*, "We don't use bullshit science. The only thing we really do is condense the time amount that it takes to get the actual results, and the other thing we do, that we cheat terribly on, is how much *money* it costs. Because we do stuff on *CSI* that nobody in their right mind would budget for the cases that we do it on. I was talking to firearms guys out in Las Vegas: 'You know that cell phone thing, where you find out who's been calling [somebody], and how long? [The phone companies] don't do that for free.'"

Hollywood Box: Story Conceits in Science Fiction

Scientists and engineers often evaluate a science fiction show or book on the quality of the science, but in doing so they forget it must first and foremost be a story, a tale that intrigues readers from beginning to end.

Below are the thoughts captured from two science fiction storytellers. One is currently in his creative prime, while the other was looking back over a lifetime of work. They are separated in time by 80 years of scientific and technological progress, but both describe remarkably similar viewpoints about how to stop a premise from overwhelming the story upon which it is built.

Kevin Murphy

Executive Producer/Showrunner: *Caprica* and *Defiance*

There's the old adage that if you boil a frog, the frog doesn't realize it's being boiled until it's too late to jump out of the pot. A science fiction audience is often like that. If you deluge your audience with outlandish/ridiculous/impossible things from Moment One, people will check out and feel, "OK, there's no point of access for me in this particular form of entertainment." But if you ask them, "OK, just stick with me for this one interesting premise," and they get hooked, then the next episode you say, "OK, now here's another layer that's also kind of fantastic and amazing...."

I'll use an example: I was the world's most gigantic *Battlestar Galactica* fan. The world of *Galactica* was very very grounded. When I was watching the original miniseries, I had the experience of going, "Wow, this is 9/11! We all got complacent, and this enemy that we underestimated struck while nobody was paying attention." But the one "gimme" that you had to buy from *Battlestar Galactica* was that instead of it being attacked by Al Qaida, it was robots ... and the robots look like us.

If you'd loaded the *Battlestar Galactica* miniseries with all of the crazy metaphysical stuff the series eventually got to—like what the frak was Starbuck, an angel or whatever—people would have checked out: "OK, this show is weak. This has no relation to my life." But, very wisely, Ron [Moore] and the producers of *Galactica*, parsed those elements out, so that as a loyal, excited, invested audience member, I was a boiled frog.

It wasn't until I got deep into the show and I was already invested in the mythology, that I understood the enormous scope of the mythology. And I think you can apply that to the science in the science fiction, to any kind of speculative fiction (Fig. 2A.1).

Herbert George Wells

Science Fiction Author (From the Preface to *The Scientific Romances of H.G. Wells*, Published in 1933)

In all this type of story the living interest lies in their non-fantastic elements and not in the invention itself. They are appeals for human sympathy quite as much as any "sympathetic" novel, and the fantastic element, the strange property or the strange world, is used only to throw up and intensify our natural reactions of wonder, fear or perplexity. The invention is nothing in itself and when this kind of thing is attempted by clumsy writers who do not understand this elementary principle nothing could be conceived more silly and extravagant. Anyone can invent human beings inside out or worlds like dumb-bells or a gravitation that repels. The thing that

Fig. 2A.1 Writer/producer Kevin Murphy and H. G. Wells. Actually, that is actor Jaime Murray, who played H. G. Wells on SyFy Channel's *Warehouse 13* … long story. Murray also plays Stahma Tarr on SyFy's *Defiance*. (Photo by Kevin Grazier)

makes such imaginations interesting is their translation into commonplace terms and a rigid exclusion of other marvels from the story. Then it becomes human. How would you feel and what might not happen to you, is the typical question, if for instance pigs could fly and one came rocketing over a hedge at you? How would you feel and what might not happen to you if suddenly you were changed into an ass and couldn't tell anyone about it? Or if you became invisible?

For the writer of fantastic stories to help the reader to play the game properly, he must help him in every possible unobtrusive way to domesticate the impossible hypothesis. He must trick him into an unwary concession to some plausible assumption and get on with his story while the illusion holds.

Hitherto, except in exploration fantasies, the fantastic element was brought in by magic. *Frankenstein* even, used some jiggery-pokery magic to animate his artificial monster. There was trouble about the thing's soul. But by the end of last century it had become difficult to squeeze even a monetary belief out of magic any longer. It occurred to me that instead of the usual interview with the devil or a magician, an ingenious use of scientific patter might with advantage be substituted. That was no great discovery. I simply brought the fetish stuff up to date, and made it as near actual theory as possible.

As soon as the magic trick has been done the whole business of the fantasy writer is to keep everything else human and real. Touches of prosaic detail are imperative and a rigorous adherence to the hypothesis. Any extra fantasy outside the cardinal assumption immediately gives a touch of irresponsible silliness to the invention. So soon as the hypothesis is launched the whole interest becomes the interest of looking at human feelings and human ways, from the new angle that has been acquired.

Unlike literary writers, who can write directly for their audience's consumption, Hollywood writers create scripts for producers, directors, production designers, and actors. Movies and shows require mobilizing large but finite pools of labor, run for a constrained amount of time, and must meet other format demands, such as a storytelling structure that works with multiple act breaks for television commercials. These concerns often have a strong influence on what conceits a Hollywood writer can and will use.

Take *Star Trek*'s iconic transporter technology. Bormanis, science consultant on three of the *Star Trek* series, explains its origin as a storytelling tool, "Gene Roddenberry wanted to do a show about a crew on a starship that's traveling the Galaxy having all of these adventures. So he said, 'I've got a space ship with a crew of 430 people, I can't land that on a planet every week, the special effects budget would be eaten up by just that. Plus we'd have to see that sequence, and we'd have to talk about this ship landing, and then see it on the planet. I can't afford that. Ah! Teleportation! I'll have this thing called the transporter. Gets our characters right into the thick of things in seconds.'"

Conceits should not be confused with two other storytelling devices that can appear superficially similar, *MacGuffins* and *tropes*. A MacGuffin is a plot device—an object or a goal—that motivates the protagonist and/or the antagonist. (The term *MacGuffin* was popularized by Alfred Hitchcock.[45]) Most often used in thrillers and action/adventure films, characters are usually willing to do almost anything to obtain the MacGuffin, but there is often little explanation as to why the object or goal is so important—the microfilm may contain the secrets of the nation, but the audience will never learn them. Often a MacGuffin is a prime focus early on to help launch the story, but the actual nature of the MacGuffin is of little overall importance and may even be forgotten by the end.

A good example of a MacGuffin can be found in the 1994 movie *Stargate*. Shortly after the beginning of the movie, the titular stargate is used to create a wormhole that transports—and then strands—the protagonists on a distant world, where they discover they must defeat an evil alien overlord. Despite giving the movie its name, the stargate doesn't make another appearance until the concluding scenes. The conceit of the movie is "What if the ancient gods were actually alien parasites who wanted humans as slave labor and hosts?" One can easily imagine an almost identical film, with the same conceit, in which the transportation technology of choice was a spaceship, matter transmitter, or even time machine.

---

[45] Although Hitchcock is often given credit for coining the term, he neither invented the term nor the concept, but rather popularized both.

In contrast, in the *Stargate SG-1* (1997–2007) television series that followed the movie, the history and operational details of the stargate device, and the physical nature of the wormholes it created, became increasingly important to plotlines as the series progressed. Very quickly, a viewer could no longer imagine swapping out the stargate for another speculative transportation technology and still have a recognizably similar show. The stargate had been promoted from a MacGuffin to a conceit.

Tropes, on the other hand, are the conceits of yesteryear, a rising tide of once incredible ideas made commonplace by use[46]. For example, once upon a time, the idea of a robot servant might have carried a story all by itself, but today seeing a few machine helpers scuttle around in the opening scenes of a movie can be a convenient way for a director to establish the setting as that of a high-tech future, confident that the audience can immediately grasp the nature and role of the metal humanoids walking the dogs, babysitting the children, or serving the champagne. That's not to say that tropes are simply window dressing—how well writers use them is often the difference between a compelling example of world-building or a messy pile of boring clichés— but they are not the drivers of a story in the way that a conceit is. Which is why, when *Stargate SG-1* spun off the *Stargate Atlantis* series (2005–2009), a new conceit was needed—after seven seasons of *SG-1*, viewers were pretty familiar with stargates. So for *Atlantis*, the wormhole technology became a background trope, while the conceit was, "what if the legendary lost city of Atlantis was not only real, but in fact a giant city-sized spaceship currently situated in another galaxy?"

Another example of a common trope, often used by screenwriters to establish quickly that a character is a scientist, is the remote or exotic laboratory. How often have we seen a scientist, mad or otherwise, introduced while at work in an underground laboratory (*Stargate SG-1, Falling Skies*[47]), in a desert laboratory (*Tarantula, The Core, Prey*[48]), in an *underground* desert laboratory (*Andromeda Strain, Independence Day, Hulk*), in an island laboratory (*Jurassic Park* and sequels), in a polar laboratory (*The Thing, Helix*), or in any place other than a research university or government laboratory, where most scientists in the real world toil. As confused as the science was throughout *Chain Reaction* (1996), the laboratory in which the hero scientists worked was an

---

[46] Not to mention the progress of science and technology in turning science fiction into science fact, such as a global computer network, ever cheaper genome sequencing, or space tourism.

[47] The character Dr. Roger Kadar, introduced working underground in season three, is loosely based upon author KRG. Originally Kadar was to have been given the nickname "The Rat King" by local children, due to his affinity for pet rats, but folks at the network were uncomfortable, thought the whole "rats as pets" aspect was unbelievable (though Skitters and Mechs and Harnesses and Espheni, apparently, aren't), and that aspect of the character was dropped. Seriously, Google "Falling Skies" and "Rat King."

[48] OK, so Michael Crichton's novel *Prey* isn't a film yet, but surely it's only a matter of time.

excellent depiction (admittedly, it wasn't right on campus, but anybody who thinks its unusual for universities to put departments that could potentially go boom at some remove should look at a map and see where JPL is relative to Caltech).

An ever-present, but scientifically inaccurate, trope is the depiction of sound transmitted through the vacuum of space: the explosion of the moon Praxis[49], the roar of a TIE fighter (complete with Doppler shift as it passes through the field of view[50])[51], the "pew-pew-pew" of Cylon and Viper laser blasters[52]. Although light waves freely pass through the void, sound, or *acoustic*, waves require a medium, some form of matter, to propagate—a medium absent in the vacuum of space.

While there are examples where TV shows (*Firefly*, 2002) and films (*2001: A Space Oddysey*, 1969) have opted for scientific accuracy on this point, it turns out that soundless explosions are so counterintuitive that depicting them as such tends to pull viewers out of the narrative every bit as much as the most heinous of "Oh, Please" moments. At a 2012 San Diego Comic-Con panel on how science is portrayed onscreen in Hollywood, when the subject of sound in space was brought up, screenwriter extraordinaire Jane Espenson offered an option that might strike the best balance between science purity and audience appreciation: "Perhaps we should just consider explosions in space part of the [music] soundtrack?"

Whether writing for the page or screen, wise scribes carefully ration the number of conceits they use. Storytellers get only so much goodwill from the audience, and their suspension of disbelief will eventually run out if story conceits are too numerous or too far-fetched (Fig. 2.5).

In contrast, a handy thing about tropes is that Hollywood storytellers can use almost any number of tropes for free without exceeding the audience BS limit and damaging viewer suspension of disbelief, even if they are wrong scientifically. Tropes like force fields, sound in space, artificial gravity, even faster-than-light travel are no less alien to science than the scorpion men of Gilgamesh[53], but over time they have become acceptably inaccurate for most viewers of science fiction.

Irrespective of the type of drama, and irrespective of the veracity of the science, there is one rule that all storytellers—cinematic, television, and lit-

---

[49] *Star Trek VI: The Undiscovered Country* (1991).
[50] See also *Enemy Mine* (1985) for a prime example of sound in space and Doppler shift from a moving spacecraft.
[51] *Star Wars* Episodes IV and greater, 1977–.
[52] Classic BSG, of course.
[53] Seriously, if scientists can splice genes from jellyfish into pigs to make porcine-shaped bioluminescent nightlights, a feat which scientists have actually pulled off, then perhaps scorpion men are an even less far-fetched creation than artificial gravity.

32   Hollyweird Science

Fig. 2.5 "Science of Science Fiction" panel at the 2014 San Diego Comic-Con. *Front to Back*: Panelists Kevin Grazier, Phil Plait, Jaime Paglia, Nicole Perlman, Jessica Cail, Andrea Letamendi, Ashley Miller, and Zack Stentz. (Photo by Stephen Cass)

erary—must follow. Once the conceits are introduced, once the setting has been established, the storyteller must follow the rules that he or she has set into motion. Tom DeSanto elaborates, "I do believe that story always has to take precedence. You've got to focus on what works best for the story. But, you can not go so far out in the logic. Once you set the rules and the logic, if you weave a story which violates that logic and those rules, the audience is going to call bullshit, as well they should. So you can't be lazy when you're writing."

It's by adhering to internal consistency that writers can overcome the inevitable discrepancies with external reality. Things that might otherwise elicit an "Oh, Please" will come across as reasonable if they fit in with the rules the story has established for itself. Kevin Murphy, executive producer for the series *Caprica* and *Defiance* says, "What's important is verisimilitude. The science doesn't need to be accurate; the science needs to be believable, because you're never going to get the science *right* … The world needs to be immersive, and you need to believe that the science is real … the science has to be accurate enough to serve the story, and to convince the layman that what's going on in the story could actually happen, and you have to believe in the world. If someone says, 'OK, I don't believe that that's properly grounded, I don't feel that they did their research, I feel that the writing has been lazy … people check out of the experience."

> Verisimilitude in a narrative is more important than veracity.
>
> Michael Crichton, Physician/Author/Screenwriter

2.4 Comparative Thresholds: The Point of Know Return

Apart from viewer enjoyment, is there any other reason we should care about the quality of screen science? We live in times when topics such as genetics, earthquakes, stem cell research, computation, climate change, space travel, robotics, and nanotechnology appear regularly in the news (and indeed much of this book will be devoted discussing many of these developments). These topics have also started to appear on ballot initiatives and in campaign debates. Our automobiles, phones, and homes are now routinely equipped with enough computational power to place them in the realm of the supercomputers of the not-so-distant past. The ability for people to understand science-related news, and its corresponding implications, looks to be an ever-important skill in the coming years. It certainly influences everybody's ability to be an informed voter, and those who don't understand basic science may find themselves increasingly disenfranchised.

So don't Hollywood productions then have a civic duty to present science as accurately as possible? Otherwise, aren't they doing a clear disservice to society, right? As tempting as it is to respond with a rousing, "So Say We All!"—and many scientists and engineers do—the hows, whys, and outcomes of science's portrayal on screen is not clear cut.

Marty Kaplan says, "I don't think people look to shows for information and facts, unless they are in a genre that is supposed to be factual, like documentaries or news. They look to shows to be entertained, and they know that entertainers are, at heart, liars. They create fiction. So when people are absorbing entertainment, the contract is not 'Teach me. Inform me, Explain to me,' the contract is, 'Don't bore me. Hold my attention. Keep me enthralled. Tell me a story.' And so that's why people are paying attention. Whether or not what they are also getting is accurate is not part of the deal."

Many television viewers simply don't pay close attention or don't care if their motion picture science is correct unless their attention is drawn to it in some way: technospeak is technospeak. So "I'll check the alignment of the ion beam in the mass spectrometer," and "I'll reverse the polarity of the neutron flow"[54] sound equally credible, and as long as the storytellers weave a compelling narrative, they're along for the ride.

Even among viewers who insist on a higher level of fidelity in order to remain engaged in a production, there are different thresholds of fidelity in what they will accept as "real" science. In his book *Lab Coats in Hollywood*, David Kirby talks about three kinds of real science: High School, Expert, and the Unknown. High school science is what a well-educated layperson will know about the universe—you can't breathe in space, our DNA contains the genetic code that dictates how our cells and bodies develop and function, nuclear explosions can destroy cities, and so on. This is the level that writers really have to pay close attention to: if they screw up here, word-of-mouth will be that their work is ridiculous, which is not likely to help the financial bottom line.

Expert level science is the material known by those with a specialist education—the planets' orbits around the sun are elliptical, multicellular life is a relatively new evolutionary arrangement on Earth, alpha radiation can usually be blocked by a piece of paper or clothing. This is the level that you can get wrong, and vast majority of your audience won't notice (though they might be less than thrilled to read about it online afterwards, and happy to join in for the subsequent finger pointing and laughing.)

Before we move onto the Unknown, we'd like to insert another level into Kirby's scheme. Increasingly, there seems to be a level of science knowledge between Kirby's "high school" and "expert" categories in play. The rise of the Internet, and in particular social media and bloggers, has spawned another level of fan, needing a different level of science veracity, which we'll call "prosumers." In addition to simply consuming products, in this case TV/film productions, prosumers tend to be active voices on social media. When prosumers can reach thousands of individuals now with a single tweet, Facebook status update, or blog post, "word of mouth," operates on a different scale—by several orders of magnitude—than it did twenty years ago, and studios and creators now actively court prosumers. A large fraction of the consumers of fictional science series have science backgrounds intermediate between those of the "high school" level and "expert" level, and if they feel that a show is smart, and respects their intelligence, they can be a show's most vocal supporters. They can also be noisy detractors if they feel a show is stupid, so wise screenwriters and producers pay them due attention.

---

[54] Jon Pertwee, the third actor to play the main role in *Doctor Who*, had a difficult time with the technical terms in scripts, but he could remember, "…reverse the polarity of the neutron flow." So he did that. Often. That phrase still makes its way into episodes even today.

Finally, there's the Unknown: speculative science. In this arena, an honest scientist should acknowledge that a writer's speculations are as valid as that of any expert, and a science advisor's creative freedom is equal to that of a screenwriter. In fact, precisely because writers are often willing to speculate where experts may be reluctant to venture beyond what is defensible with known facts, science fiction has gained some respect among policy makers grappling with the implications of rapid technological development. It's not unknown for science fiction authors to be called to Washington D.C. to give their two cents to various agencies or legislative committees concerned with issues such as homeland security, climate change, or space exploration. Science fiction novelist David Brin, who has been on the receiving end of several such summons, elaborates:

> I believe that a civilization that pays heed to high-end science fiction is expressing basic logic and wisdom. One can look at a track record of successful predictions, or vital warnings, across the last 60 years. Folks like Vernor Vinge, Stan Robinson, Joe Haldeman, Greg Bear, peer into tomorrow for possibilities and probabilities and mistakes. These stories we create aren't the result of vast analytics. These agencies and corporations spend most of their attention gathering huge amounts of data and figuring out new ways to massage them, but to spend a few hours per year probing our trained imaginations would seem to be worth the time and effort.

For similar reasons, author SAC has edited anthologies of literary science fiction for the otherwise fairly hard-nosed technology publications *MIT Technology Review* and *IEEE Spectrum*. As loudly as some will complain about an "Oh, Please" moment created by a lazy screenwriter, it's equally unimpressive when an unimaginative scientist condemns a story as being bad science, when it is, in fact, a legitimate venture into the unknown. Brin shares, "Science fiction, in particular, is strongly associated with creativity, and a willingness to explore both good and bad outcomes of change. I believe, that a culture's strength and resilience is measured, in some degree, of the health and vigor and enthusiasm of its science fiction."

2.5 Hollywood Science and Science Education— The Condensed Matter Version

Count science educators in the category of people who certainly have concerns about scientific accuracy in Hollywood productions, and research supports that there is valid reason for these concerns. The Norman Lear Center at

of Southern California is a think tank that "studies the social,
omic and cultural impact of entertainment on the world." Its
artin Kaplan, elaborates that the mission of the Norman Lear
Center, is to study and shape the impact of media and entertainment on
society: "On the studying front, we do research—both qualitative and quantitative research—on media impact. On the shape front, we are meddlers, in-
terveners, advocates, and do-gooders, trying to use media and entertainment
to make the world a better place."

Kaplan, who has described Hollywood productions as the "unofficial curriculum of society," elaborates:

> In Plato's time, the entertainment industry, the rock stars, were the poets who
> recited Homer, who had audiences of tens of thousands of people, and who
> competed for huge cash prizes. Plato asked, "In an ideal state, what should
> be the *curriculum* of our philosopher-kings? And he said, "Geometry, and
> arithmetic, and all the other sciences, but we can't let them be exposed to
> entertainment, to poetry. Because, no matter how smart they are, no matter
> how rationally trained, no matter how much they realize that it's 'only' poetry,
> it will affect them." Poetry, entertainment, has access to a different system of
> the body. It bypasses the reason, and even if you think you know better, it will
> affect you and change you. When we are in that state, stuff we are seeing gets
> access to a part of us that is not under our conscious control. It affects what we
> know, and what we believe, and how we behave. That's whether or not that's
> intended by the people who create the entertainment, and whether or not it's
> intended by the people receiving it.

Just as anger is the fast path to the Dark Side of the Force, content presented
as entertainment seductively allows inaccurate scientific concepts easy access to our brains. In support of this, a 2006 paper by Michael Barrett from
Boston College together with colleagues, published in the *Journal of Science
Education and Technology*[55], found that students who had watched the film
The Core had several misunderstandings of Earth science concepts that were
not shared by students who hadn't watched the film. Barnett and colleagues
reported, "… the film leveraged the scientific authority of the main character,
coupled with scientifically correct explanations of *some* basic Earth science, to
create a series of plausible, albeit unscientific, ideas that made sense to students." Although their sample sizes were small, this study provides evidence
of an instance where entertainment paved the way for inaccurate science to be
taken as fact for the research subjects.

---

[55] Barrett M, Wagner H, Gatling A, Anderson J, Houle M, Kafka A (2006) *Journal of Science Education and Technology* 15 (2), pp 179–191.

Almost everyone shuts down when science becomes too technical; you've got to infuse it with entertainment and storytelling to make it effective.

Greg Graffin, Musician/Science Educator

Neuroscientists are beginning to discover the scientific basis for this. Several studies using functional magnetic resonance imaging (fMRI) have uncovered that the human brain makes no distinction between reading about an experience and encountering that experience in real life. When test subjects read words corresponding to real-world items that have a particular smell, like "coffee," "perfume," or "cinnamon," not only did the region of the brain associated with word recognition activate, so did the region of the brain associated with scent processing. When subjects that read words and phrases associated with motion, the region of the brain that coordinates the body's motions activates. Words that described textures activated the region of the brain that received tactile stimuli. Succinctly, the brain does not distinguish between information it receives from within a fictional narrative, or gathered from real life. If you doubt this, simply check your Facebook news feed or Twitter feed immediately after a major character dies on *Game of Thrones* or *The Walking Dead*.

If science fiction tropes like "We only use ten percent of our brain, just think what we could do if we used it all" are recurring topics in TV and film[56,57], the implication is that scientifically inaccurate concepts are making their way from the screen into the minds of viewers, and internalized as fact. Writers of TV and film tend also to be rabid connoisseurs of TV and film, often from a young age, so as they are exposed to these tropes they repeat the cycle[58]. There is, in fact, a new field of study called *agnotology* that studies culturally induced doubt or ignorance, with special emphasis on inaccurate, misleading, or cherry-picked scientific data. Oft-repeated science inaccuracies in TV/film are, then, highly efficient modes of disseminating doubt and ignorance, and we will call this feedback loop—where inaccurate science is internalized by a (future) screenwriter and repeated as fact—the *Hollywood Curriculum Cycle*. It will rear its ugly head again in later chapters.

---

[56] Neuropsychologist Donald McBurney called the Ten Percent Myth, "One of the hardiest weeds in the garden of psychology."
[57] On the TVTropes web site, there is an entry which states, "Eureka's scientific adviser was on record as making sure that when that concept was used on the show, the line was 'ten percent... at any one time' specifically to dodge this trope." It's nice when your work is appreciated. http://tvtropes.org/pmwiki/pmwiki.php/Main/NinetyPercentOfYourBrain.
[58] *Lucy* (2014) Enough already!

A very similar situation can occur when a screenwriter, with every intentions of incorporating science that's as accurate as possible, looks to the Internet. One trait that often separates good screenwriters from mediocre ones is a willingness to do the necessary research for a script, but even motivated writers with the best of intentions sometimes simply do not have the education or background necessary to discern what is science, what is dubious science, and what is pseudo-science. Much of the junk on the Internet comes wrapped in seductive packaging. Even back in 2000, the National Science Foundation, in their yearly *Science and Engineering Indicators* observed, "The amount of information now available can be overwhelming and seems to be increasing exponentially. This has led to 'information pollution,' which includes the presentation of fiction as fact. Thus, being able to distinguish fact from fiction has become just as important as knowing what is true and what is not."[59]

In an odd twist, some television series that normally have a very high regard for scientific fidelity, the science is, on occasion, made inaccurate intentionally. If entertainment is an efficient vehicle for delivering information to our brains, there are some forms of science content best rendered inaccurate before being broadcast. On crime procedurals like *CSI*, writers and their science consultants research the design and construction of various lethal devices and then, as writer Bradley Thompson says, "We spent a lot of time on *CSI* finding out what *wouldn't* work, so that we could put that in the show, because we didn't want to be teaching people how to build bombs on television."

A similar situation occurred on the series *Breaking Bad* (2008–2013). Donna Nelson, chemist and science advisor for the series, explains how the series avoided being a cooking show for methamphetamine addicts: "Producer Vince Gilligan told me that he sought assistance from the DEA on how to present syntheses of methamphetamine in a manner that would be both scientifically accurate and yet safe for public viewing. Two common methods were used to accomplish this: one, a critical step in a synthesis would be omitted, or, two, steps from multiple syntheses would be combined. All the chemical transformations shown were accurate, but they would not be presented in an order that would produce meth."

A second plot twist is that the same ubiquity that makes a trope a trope also provides a teaching opportunity. Any stand-up comic would agree that no audience member is more engaged with your performance than the heckler. While the trend in Hollywood is to get science increasingly accurate, it will never be perfect. Rather than complain about inaccuracies, educators will find that there is bountiful opportunity when screen science—even poorly

---

[59] National Science Foundation (2000) *Indicators: Science and Engineering 2000*. National Science Foundation, Washington, DC.

done screen science—can be heckled to create entry points into discussion of real science. There is an expansive body of literature covering how science portrayed on TV and in film, both well and poorly, can be used for science education, and we hope in this book to do just that.

Ironically, given frequent complaints about scientific errors in screenplays, it's worth noting that science educators and science curricula developers have always been willing to fudge science to elucidate difficult concepts themselves. The concept of *models*—highly oversimplified representations used for pedagogical clarity—dates back to circa 500 B.C. with Thales of Miletus, often called the "Father of Science"[60]. For example, the orrery—the simplified model of the Sun and planets that sat gathering dust in your elementary school library—is a type of model used to depict the sequence of the planets from the center of the Solar System, as well as the fact that they all orbit the Sun. It ignores the fact that planets don't move in perfect circles, but in somewhat elliptical orbits. The proportional distances between the planets are also usually incorrect: the outer planets are in fact so far apart, then when Uranus was discovered in 1781 it *doubled* the diameter of the known Solar System.

The Bohr model of the atom is another common simplified model. The atomic representation taught to high school students, where electrons are dots strung along neat circular orbits around a nucleus[61], is a far cry from the three-dimensional probabilistic sculptures that undergraduate chemistry and physics students learn.

It is not that such models are not useful educational tools—we will be using them ourselves (e.g., in Chapter Four, "Matter Matters")—they are extremely beneficial, and there's a reason that simplified models have been used in science education for 2500 years. When science educators complain about TV/film inaccuracy, though, they are, intentionally or otherwise, reserving for themselves the privilege to be inaccurate when it suits their needs, justifying this double standard on the basis that the needs of science educators are pure, they know what they're doing, and Hollywood is just … Hollywood. We believe the situation is not so black and white.

Pedagogical "cheats" are, in fact, used in Hollywood as well. Each episode of the highly popular[62] children's science education cartoon *The Zula Patrol* (2005–2010), focused narrowly on a featured science topic. Other aspects and topics within the realm of science—even if they had been the focus of

---

[60] Although some argue that Democritus deserves that Moniker, we've already taken the reader through science fiction's "Who's yer daddy?" arguments, we'll spare you a second one.

[61] An excellent example of this is the scene wipe graphic used in *Big Bang Theory*.

[62] When *The Zula Patrol* moved from PBS to NBC, it became one of the most-watched shows on daytime television. Although targeted for children aged 4–7, for a while it was the most popular show in the female 18–49 demographic – among women who watched the show with their children.

—were allowed to "drift." While an effective tactic for deal-
ured topic at hand, the potential for confusion within the
s clear if viewers recall previous episodes, and can note the
n when a science topic was the focus, and when it was not.

nagined documentary series *Cosmos: A SpaceTime Odyssey*
(2014) has, at times, opted for artistic license rather than strict adherence to
reality. The series is told from the vantage point of a craft called the "Ship of
the Imagination," a clever narrative conceit that allows the show to examine
every aspect of the Universe, from the very big to the very small to the very
dangerous.

Within the first 10 minutes of the pilot episode, the Ship of the Imagination is depicted traversing a densely populated asteroid belt. The "jumble
of space rocks"—like the ones Han Solo dodged in *The Empire Strikes Back*
(1980)—is an inaccurate trope that has been repeated time and again, and
has gathered the ire of planetary scientists and astronomers time and again[63].
Although asteroids can, on very rare occasion, collide[64], most of the time it
would be impossible to see one asteroid in the asteroid belt from another (with
the exception of binary pairs of asteroids like Ida and its "moon" Dactyl). Even
more inaccurately, a scant two minutes later, *Cosmos's* Ship of the Imagination
is depicted traversing a similarly dense debris field at the edge of the Solar
System: the Kuiper Belt[65]. This is an even more sparsely populated belt of icy/
rocky objects orbiting the Sun beyond the planet of Neptune, and here the depiction was even less accurate. Are screenwriters allowed such "artistic license"
in a series labeled as a "documentary"? The jury is still out on this (Fig. 2.6).

The same "information overload" that makes researching science difficult for screenwriters also muddies the waters for viewers, since Hollywood's
fictional science presentations become part of that overload. Moreover, how
do we collectively filter less-than-accurate science conveyed within the ultra-
efficient delivery system that is entertainment? Short of a scientific or educational "rating system," the answers aren't clear.

> False facts are highly injurious to the progress of science, for they often endure
> long.
>
> Charles Darwin

---

[63] The field of tumbling space rocks in the fan-favorite *BSG* episode "Scar" was intended to be a protoplanetary disk rather than an asteroid belt. This distinction does not make the depiction correct, but
does make it less wrong.
[64] Analysis of mineral fragments of the bolide that exploded over Chelyabinsk, Russia in 2013 strongly
suggest that the object was placed on an Earth-crossing trajectory as the result of an asteroid-asteroid
collision roughly 300 million years ago.
[65] Also known, perhaps more accurately, as the Edgeworth-Kuiper Belt.

Fig. 2.6 Ida and its moonlet Dactyl. (Image courtesy of NASA/JPL-Caltech)

2.5.1 Return to Nerdvana

Hollywood productions can be an efficient vehicle for delivering accurate science to minds made receptive by being in "entertainment mode." These productions can also serve to propagate inaccurate science for the very same reason. Production concerns like cost and schedule notwithstanding, the level of science fidelity that a production adopts is a very thorny matter. The American audience has a hunger for science accuracy; the public discourse in the Twitterverse and blogspheres in October 2013 debating the accuracy of the film *Gravity* (2013) reflects this.

If there are dollar signs attached to improved scientific accuracy, then rest assured that improved science accuracy is what will happen, since a key goal of any Hollywood production is to attract and retain as large of an audience as possible. Clearly there are cases, however, when rigid devotion to scientific truth is impossible, unnecessary, or unwanted[66]. The choices to make the science as accurate as possible, cheat a bit and make it *almost* accurate, or abandon accurate science entirely, can have a very real impact on audience enjoyment and appreciation—as well as the financial bottom line—making the accuracy issue even more convoluted.

> When we're under the spell of entertainment—think of the words, we are enthralled, entranced by entertainment—it is a different state of being.
>
> Martin Kaplan, Ph.D., Director, Norman Lear Center, USC Anneberg School of Communications.

---

[66] Rephrased: Marvel, keep doing what you're doing! We're big fans.

The path to Nerdvana—the state where a viewer can enjoy a science-themed production and not be pulled out of the story because of some aspect of the science—is both a complicated and highly subjective one: the path is different for every viewer. For the creators of Hollywood science fiction and science-themed programming, there is no Yellow Brick Road to Engagement City, but rather a twisted wibbly-wobby sciency-wiency labyrinth straight out of an Escher print.

Science Box: Turning a Trope into a Grope—Just Add G.

Just as tropes are commonly occurring narrative devices or clichés in TV/film, there are also commonly recurring complaints that appear on the Internet moments after an episode or film has aired—low-hanging complaint tropes, if you will. A gripe that is also a trope[67], we refer to a *grope*[68].

One of the more common gropes occurs when a drama depicts space-based warfare—specifically dogfights—and fighters carve banked turns in space. Internet gropes are frequently laced with condescension ("When I saw this on *BSG*, I first though [sic] 'The writers are smarter then [sic] that! Why are spaceships banking??'"), self-importance ("The ultimate in bad tech had to be *Battlestar* [sic] *Galactica*... ships that ROARED through the vacuum of space and made banked turns as tho [sic] in air...."), and snobbery ("Even Kevin Grazier would have to agree that it is a physical impossibility to do banked turns in space"). What these complaints share is that they assume that the issue is entirely aerodynamic, and they neglect one very important component of the equation: the pilot.

In Fig. 2B.1, we see our test spacecraft, the *Century Parakeet*, performing a turn that creates a positive G[69] for the pilot. This is the kind of turn that happens when a pilot pulls back on the control stick, lifting the nose of the craft. The vector F_p represents a positive G, indicating the direction of the force on the pilot due to the turn. This is equivalent in direction (though possibly different in magnitude and would be present even in space) to the force of gravitation forcing a pilot down into a chair

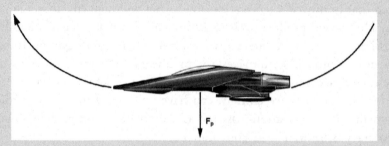

Fig. 2B.1   A fighter pulling a positive G turn. (Illustration by Eric Chu)

---

[67] One that typically winds up in a comment thread on the Internet.

[68] Appropriate, because oftentimes people who go to this well are simply groping for *something* to complain about.

[69] One G is the equivalent to the force of gravitation at Earth's surface. Two Gs are twice that force, and so on.

on Earth, and although this type of G force can be physically demanding—depending upon how fast the craft is travelling and how tight the pilot turns—it is the easiest form of G force for a pilot to withstand (**G** force is usually referred to as **g** force in other fields, and elsewhere in this book, but the capital **G** is the form used by aviators.).

Hollyweird Science spoke to Major Christina Szasz, who has flown combat missions in the F-16 and the F-117 aircraft, and asked her to explain the forces a pilot experiences in a turn, and what pilots do to compensate. During a high positive G turn, just as the pilot is forced down into the seat, blood is forced downwards as well, pooling at the pilot's feet. Enough blood loss from the head can cause a pilot to experience *G-LOC* (G-force Loss of Consciousness). Major Szasz explains how a pilot fights against that, "We are trained to do a G-strain. You start by tensing your calves, then move up to your thighs, your quads, and your glutes, then your abdomen. You have to clench it in that order, from the bottom up. You can easily G-LOC if you don't use a G-strain properly."

Fighter pilots also have a G-suit that helps a pilot cope with high-G turns. Szasz continues, "It's a garment that you put around your legs and also around your abdomen, and there is a hose attached to it. Inside of the fabric are bladders. You plug the hose into the aircraft, and as the aircraft senses a certain amount of Gs being pulled, it regulates how much air fills the bladders to increase the pressure against your muscles and your skin. The purpose of that is to compress your muscles, and help your body keep the blood elevated up closer to your heart. It only works in conjunction with the G-strain. Meanwhile, the more Gs you pull, the more air is filling the bladders, which increases the pressure to help press the blood to the upper part of the body. In the sophisticated fighters of today, and presumably those of the distant future[70], Szasz says that even with a G-suit and techniques to fight the effects of high-G turns, "Typically the aircraft can handle much higher Gs than the pilot can, that's why they limit the F-16 to 9 Gs, because that's the max for the human body."

Figure 2B.2 depicts the pilot pulling a negative G, with vector F_n indicating the direction of the force acting on the pilot. This is the type of turn that happens when a pilot pushes forward on the control stick, dropping the nose of the craft, and producing the same kind of feeling as that first long drop on a roller coaster.

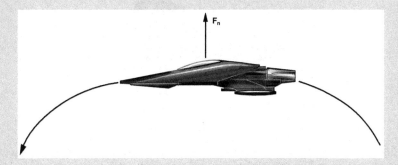

Fig. 2B.2 Negative G. This turn forces fluids—like blood and the pilot's lunch—to move vertically towards the *top* of the craft. (Illustration by Eric Chu)

---

[70] If, in fact, they have pilots at all.

This type of G force is more difficult for a pilot to withstand that a positive G, especially if it is higher than one G of force. Szasz says, "A negative G is very painful, where now instead of the blood wanting to pool at your feet, it now wants to pool up at your head. That's pretty dangerous, because now you have an excess of blood rushing to your head, and that's where you get the red-out, as opposed to the grey-out[71]. In a positive G, you'll get a grey-out, which means you're losing blood from your brain and your eyes and your vision, whereas in a negative G a lot of blood is being pushed up into the head."

While it is tempting to believe that the red drowning out the field of view is due to blood being forced into the eye tissue, more likely it is caused by the pilot's lower eyelid being forced over the eye—it is forced upwards along with everything else. Nevertheless, a red-out is a potentially dangerous, even lethal, situation. With that much blood flowing to the soft tissues of the head, a red-out can lead to eye damage, even stroke[72].

When you accidentally go into a negative G, you recognize it pretty quickly, because it's pretty painful, and so you let off. It's very hard to self-induce a negative G for a long period of time." There is nothing—no technique or hardware—to mitigate the physiological effects of negative Gs. Szasz adds that the best way to minimize the effect of negative Gs on a pilot is, "Don't do it."

What if a fighter pilot sees an adversary at a lower altitude, or is attacked by an enemy fighter, and needs to dive? Szasz explains that, no matter what, orienting the aircraft in order to pull a positive G is always preferable: "If you find yourself needing to dive, more than likely you're going to invert the plane, and pull up versus trying to push through a negative G. Pulling a negative G would actually make the situation worse, because it would take longer to recover."

Figure 2B.3 shows a flat turn, which subjects a pilot to a *transverse G*. Vector F_t shows the direction of the force—it has a tendency to slam the pilot's head against the canopy (the right side in Fig. 2B.3). This type of G force is more common for operators of craft, like automobiles[73], that are constrained to move in two dimensions.

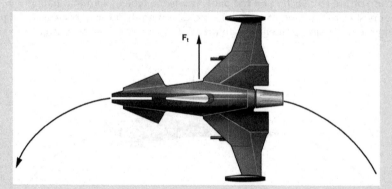

Fig. 2B.3   Transverse G. This type of G force is what most people experience when they turn their automobile while driving. It is very difficult for an aircraft to do, but possible for a craft in space. (Illustration by Eric Chu)

---

[71] Szurovy G, Goulian M (1994) *Basic Aerobatics*. McGraw-Hill, New York, pp 33–34.
[72] DeHart RL, Davis JR (eds) (2002) *Fundamentals of Aerospace Medicine*, 3rd edn. Lippincott, Williams & Wilkins, Philadelphia.
[73] Certain carnival rides as well.

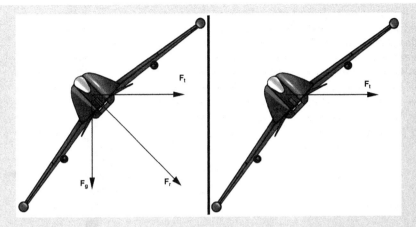

Fig. 2B.4 Banked turn with and without gravity. (Illustration by Eric Chu)

Fighter craft are generally incapable of pulling much in the way of transverse *G* forces, unless in a flat spin, a very dangerous situation[74]. Fighters in an atmosphere, instead, bank their turns in large part due to aerodynamics, but also for the pilot. The left-hand panel of Fig. 2B.4 shows the *Parakeet* in a banked turn with gravity acting upon the spacecraft. The force of gravitation is F_g, the force due to the turn is F_t, and the resultant of the two is F_r. The craft is banked such that the resultant force is a positive *G*.

The right-hand panel of Fig. 2B.4 shows the craft in the same banked turn as the left-hand panel, but in the absence of gravity. In this instance, the force the pilot feels has both positive *G* and transverse *G* components. Not only would this be an unusual direction for the pilot to feel a *G* force (down and to the side), it would take an improbable array of thrusters to achieve.

In order for the pilot to best orient the craft to withstand the *G* force of the turn, a 90° bank would be in order, turning this, and every turn, into a positive *G*-turn. In Fig. 2B.5, the *Parakeet* is banked 90° relative to its pre-turn orientation. This means that vector *Ft*, the force of the turn, is equal to F_p: a positive *G*.

So the problem in films like *Star Wars* and TV series like the original *Battlestar Galactica* is not that the spacecraft carve banked turns, it is that they are not banked *enough*.

The fact is that if there were space dogfighting craft, these types of issues with *G* forces would likely be an issue for the first generation only (if even that). It has been said that weapon systems are "built to fight the last war." Historically, new technology is often designed, or used, like high-tech versions of the previous generation of tech[75]. A real space fighter would be more likely to behave like the Starfuries depicted on *Babylon 5* or the Vipers of the reimagined *Battlestar Galactica*. The craft would use thrusters to reorient the craft to the proper attitude, and use the main thruster for any course changes. That a plausible explanation for banking turns in space is simple to envision, most complaints on the topic seem less well-thought-out, and more outright malicious.

---

[74] Recall that it was a flat spin in *Top Gun* that ultimately cost Goose his life.

[75] A notable exception today is the tank-killing USAF A-10 Thunderbolt II aircraft. Its close air support effectiveness and survivability is why the aircraft has staunch defenders desperately trying to forestall its retirement.

Not all viewers grope. Most, in fact, are fair and even relish when their favorite series gets science correct. One fan noted on a chat board, "*Battlestar* allowed ships to be moving in a direction other than the direction they were pointed." Comments like these are common as well. Too frequently, though, viewer complaints about inaccurate science are often fraught with inaccurate science.

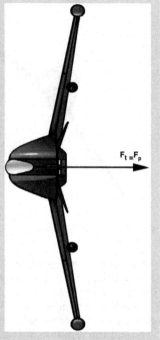

Fig. 2B.5 If the pilot banks the craft 90°, a transverse *G* becomes a positive *G*

3

Hollywood Scientists: Reel and Imaginary

It's very important for us to see that science is done by people, not just brains but whole human beings, and sometimes at great cost.
Alan Alda, Actor/Director

There is one thing even more vital to science than intelligent methods; and that is, the sincere desire to find out the truth, whatever that may be.
Charles Sanders Pierce, Scientist/Philosopher

The mind sees what it needs to see. The soul sees what the soul sees.
Soul Hunter, *Babylon 5*, "River of Souls"

If you broach the topic of how scientists are portrayed in Hollywood with scientists, prepare yourself for some big-time science friction. In a 2005 *New York Times* interview, filmmaker James Cameron (*Terminator, Titanic, Avatar*) stated that in TV and film, scientists are typically depicted as "idiosyncratic nerds or actively the villains"[1]. This oft-cited quip is a succinct reflection of the way that many scientists believe—often quite passionately—that Hollywood portrays their profession[2,3]. Poll scientists and you will find that they are convinced that, particularly in works of science fiction, Hollywood depicts them as some permutation of megalomaniacal, nerdy, scheming, or eccentric—and that's before we get to the *real* character flaws.

There is, however, a significant disconnect between this belief and a recent content analysis of primetime television by Anthony Dudo and colleagues. They showed that between the years 2000 and 2008, although scientist char-

---

[1] Revkin A (2005) Filmmaker Employs the Arts to Promote the Sciences, *New York Times*, Feb. 1
[2] Although this quote is often cited in an environment of martyrdom, the follow-up question that nobody ever seems to examine—but which we do later in the chapter—is, "Is this a fair portrayal?"
[3] From *The Abyss* to *Solaris* to *Avatar*, James Cameron's scientist characters have always been portrayed extremely well.

acters were relatively rare, when they did appear, it was almost exclusively in a positive light[4].

How have scientists'[5] perception of Hollywood's portrayal become so skewed? Why are scientists portrayed as they are? These are complex questions, but teasing out the answers provides insights into the role of both science and science fiction in society.

3.1 Scientist Representations

> Let's be clear: all professions look bad in the movies. And there's a good reason for this. Movies don't portray career paths, they conscript interesting lifestyles to serve a plot. So lawyers are all unscrupulous and doctors are all uncaring. Psychiatrists are all crazy, and politicians are all corrupt. All cops are psychopaths, and all businessmen are crooks. Even moviemakers come off badly: directors are megalomaniacs, actors are spoiled brats... why expect scientists to be treated differently?
>
> Michael Crichton, Physician/Author/Screenwriter

If Cameron is correct, Hollywood TV and cinema has been fairly consistent in casting scientist characters as nerdy except, of course, when they're cast as megalomaniacal. Dr. Marty Kaplan, Director of the Norman Lear Center, adds, "...I would amend James Cameron's comment: 'and they're men.'"

According to a 2001 study by the National Science Foundation (NSF), there were just over 400,000 Ph.D.-level scientists in the United States. Data from the 2010 U.S. Census and a 2011 report from the National Science Foundation show that, of roughly 308 million people in the United States only 1.3 percent of the population aged 25 and older have Ph.D. degrees and just over 561,000 (1 in 500) are Ph.D.-level scientists.

In the United States, natural redheads represent approximately 2 percent of the population. Look around. Count how many redheads you see. Not many? Take a longer baseline and count for a day. A week. Still not many? A number equal to about one tenth of that is how many doctorate-holding scientists are scattered throughout the American population. In some parts of the country they are even rarer as they tend to be concentrated around universities, government laboratories, and corporate tech parks[6]. The odds are high that if you

---

[4] Dudo A, Brossard D, Shanahan J, Scheufele DA, Morgan M, Signorielli N (2011) Science on Television in the 21st Century: Recent Trends in Portrayals and Their Contributions to Public Attitudes Toward Science. *Communication Research* 48(6): 754–777

[5] And science writers and James Cameron.

[6] Whereas redheads tend to be concentrated in Scotland, Ireland, and Boston.

are not a scientist, you've never met a real working scientist, w
exception of your university science class (and, then, did you
professor?). If that's the case, then it's likely that your concept c
tist represents, how a scientist behaves, and how a scientist spe
from television or film.

It is in television and film that the average viewer has "access" to the lives of
the rich, the powerful, the influential, and the interesting. Because a goal of
entertainment is to present compelling characters in interesting situations, the
demographics of the careers of the fictional characters on screen do not cor-
relate with the occurrence of those same careers in society. In 2012 there were
some 4.6 million Americans with retail jobs, while there were just 690,000
physicians and surgeons. Yet think of how many doctors we see on TV as central characters compared to retail workers. Physicians and surgeons are over-
represented on television because they are interesting, work in a fast-paced
environment, and often labor at the intersection between life and death. Similarly, scientists are interesting, work with bleeding edge technology, and push
back the forefront of human knowledge. Like physicians, they're also over-
represented on screen. Right?

Not so much.

The Dudo study did show that, on primetime television, 1 percent of characters are scientists. Based upon that number, you could claim that, with a
representation of only 0.2 percent in real society, scientists are over-represent-
ed by a factor of five. A more statistically accurate response is that because
there are very small numbers of scientists in absolute terms, the addition or
subtraction of a single character on a TV show changes the value dramatically,
and you can reasonably claim that scientist representations on prime-time
television actually closely mirrors their distribution within society.

In any case, with real-life examples scarce, what does a screenwriter draw
upon in order to depict a scientist? Here we have another manifestation of the
Hollywood Curriculum Cycle[7]—with few actual role models. For years, Hollywood was locked in a self-perpetuating feedback loop. Kaplan elaborates:
"When writers depict scientists, they probably do what they do in every other
realm, which is draw from their own experience, and whether that experience
is personal or, more likely, from other entertainment. They have seen what
"scientists" look like. It's in Frankenstein movies, and in cartoons, and that
helps give them a [reference] frame." Screenwriter Zack Stentz adds, "Yeah,
it's like all the cop shows are written by people who watch other cop shows."

The problem was compounded by the fact that there were also few examples
in television and film, causing depictions to become ever-narrower each time
through the Curriculum Cycle. Author Robert Sawyer explains, "In general,

---

[7] See Chapter Two "The Path to Nerdvana"

₁ou can have any profession at all, you know, security guard, probate lawyer, hostess at a restaurant, you name it, but you can't have somebody who is a quantum physicist or a geneticist or something like that as the main character in a TV series. It's just a non-starter."[8]

Sawyer should know. He shares his experience when his novel was developed as an ABC series: "My novel *FlashForward* is set at CERN, the European Center for Particle Physics, and the main characters are all physicists or engineers. We were told right up front [by ABC] that was a non-starter for American television: that your main characters could not be scientists because those are 'un-relatable' characters for the general public." Sawyer adds, "There has never been, that I can think of, a successful [pre *The Big Bang Theory*] non-science fiction American television series about a scientist who was not a forensic scientist or a medical scientist."

In many aspects, forensic scientists and physicians work more like practicing engineers than research scientists[9]. Both rely on high tech to do their jobs, as well as procedures based upon the scientific method, but while these types of careers are both highly challenging and necessary, their challenges are of different natures than research scientists and engineers[10,11], and they don't live at the nebulous gray boundary between the known and the unknown. The differences between these practical scientists and research scientists is what, in the minds of studios and networks, makes one cadre of scientists "relatable," and the other "unrelatable." Marty Kaplan, a former network executive with Disney, who we met in Chapter Two, explains the significance:

> If television executives believe that scientists are not relatable, then that is the cultural assumption in the community that they move through. They could also say that anyone over 29 is unrelatable[12], or that rural stories are unrelatable. Relatability is one of the holy grails of network television.

Certainly scientists in the real world share several "relatable" traits with everybody else: they use the restroom, they shop at Target, they drink wine, they go to the dentist, they have sex, they get in automobile accidents. Nevertheless, Kaplan adds, "Yes, but that studio executive doesn't know any scientists."

---

[8] The year 1971 saw the debut of the *Jimmy Stewart Show*, about a small-town professor of anthropology. Hardly successful, the show received poor reviews, and was cancelled after its first season run of 24 episodes.

[9] Ignoring, for a moment, that much of the cutting edge biomedical research done today is done by M.D.s or M.D./Ph.D.s

[10] For example, in the case of forensic scientists, there's gore, weather and the elements, gore, extreme attention to detail, analysis of every known human bodily fluid, and gore. For that matter, the medical scientists have the same list, just sans the weather and elements.

[11] Yes, we realize that we are vastly over-emphasizing the gore aspect, but a little goes a very long way.

[12] Holy *Logan's Run!*

Even those scientists who do have broad name recognition—Newton, Tesla, Einstein—do little to shift the needle, because most people are ignorant of the individual humanity of these icons. All had—and, to some degree, still have—devoted cults of personalities, and have been reduced to generic and somewhat zen-like beings[13].

Newton's work calmed a world struggling with the post-Copernican revelation that Earth is not the center of Creation or, really, anything. His seminal work, the *Philosophiæ Naturalis Principia Mathematica* (1687) implied that the Universe operates by laws that, although largely unknown, are at least knowable. Einstein's work provided the world with the most famous equation in history, $E = mc^2$, and instilled in the public the belief that a significant amount of how the Universe functions can be explained by the application of a simple equation. Tesla had yearly birthday bashes where he invited members of the media and gave technological demonstrations of his inventions that bordered on magic[14].

Newton was also an arrogant, vindictive jerk who had many professional knock-down/drag-out feuds. When Newton was master of the Royal Mint, he disguised himself so he could personally track down counterfeiters and send them to the gallows. Tesla could be harsh, and was narcissistic, incredibly obsessive-compulsive, and a proponent of eugenics. Einstein had at least half a dozen extra-marital affairs.

So with only a few superficial examples for screenwriters, what tends to happen to scientist characters? Stentz's writing partner Ashley Edward Miller explains, "I think it becomes very easy in a story for writers to look at their scientist character as a functionary, and the function that they ascribe to the scientist is explainer, exposition dumper, somebody who provides information or analysis that other characters can not, and provides it with an authority that the audience will accept. The problem with that scheme is that it dehumanizes the scientist character, it puts that character out of reach in terms of the audience's ability to truly relate."[15]

In the late 1980s' and 1990s' Hollywood, TV dramas like *Hill Street Blues* and *L.A. Law* proved the dramatic power of shows with an ensemble cast: where every character is the focus at one time or another, and instead of a solid "lead," all characters share a roughly equal amount of screen time. Series like *E.R.* (1994–2009), and even *Star Trek: The Next Generation* (1987–1994)

---

[13] Jonas Salk, credited with inventing the polio vaccine, had a celebrity following at least equal to these three scientists for a time. The key difference is that Salk did not enjoy the limelight, and he didn't have a particularly quirky personality. He was just a hard-working, dedicated scientist who felt that his celebrity status often detracted from getting real work done.

[14] Though the story about the device that Tesla held in his hand during one of these demonstrations, the one that created a local earthquake in NYC, is a tall tale: Tesla had a flair for hyperbole.

[15] A great example of this is The Professor on Gilligan's Island, who was used almost exclusively as gene-spliced explainer and MacGyver.

displayed the dramatic potential when all the characters were complex and interesting, and more than one or two are presented as technically competent. Stentz shares, "When we [Stentz and Miller] were writing for *Fringe* (2008–2013), we actually had a lot of fun in the scenes when we could have Walter bouncing ideas off of Peter, his son, who is also brilliant, and Astrid who, if it weren't for those two guys, would be the smartest person in any room. So it's interesting that hypercompetence tends to be focused in one or maybe two characters in an ensemble, when there's actually kind of great storytelling to be had by having a room full smart people. I mean look at the great problem-solving scenes in *Apollo 13*." Stentz adds, "We pushed to make the characters more plausibly scientific, and actually have them make insights that, in some cases, were based in real science and not BS pseudoscience."

This period also saw the explosion in the popularity of cable television. Scientist characters, with their potential for new and different portrayals, were mainstays on the new Sci-Fi Channel. Viewers could watch real scientists all day on the Discovery Channel, National Geographic, and even History Channel. Still it was another quirk of cable television that really opened the doors for today's new breed of scientist characters like Dexter Morgan, Walter White, even Gaius Baltar. While acknowledging that relatability is an important attribute of characters on network television, Marty Kaplan goes on to say, "By contrast, *unrelatability* has become the holy grail for cable—where repellant characters have proven to be fascinating, and get high ratings".

This period also saw the rise of the science advisor. While they had been used sporadically throughout the years—Stanley Kubrick consulted numerous scientists and engineers, and over 50 corporations[16] in the making of *2001: A Space Odyssey* (1969)—it's arguable that the seeds of this rise were sown with the creation of *Star Trek* in the 1960s. When Gene Roddenberry was first developing *Star Trek* (1966–1969), he consulted scientists and engineers at both JPL and the RAND Corporation, and the later *Star Trek* series all had a standing science advisor who could not only advise on scientific matters, but also on the portrayal of scientists. Many of these series aired over many years, employing many writers and producers over their run. When these writers and producers moved on from the *Star Trek* franchise, and became showrunners themselves, many hired their own science advisors, helping greatly to popularize the usefulness of this role in both television series and film (Fig. 3.1).

With ensemble casts and the popularity of cable, the potential for more, and better characterized, scientists on screen began to be realized. Nevertheless, the behavior and scientific competence of a screen scientist is still limited by the experience and knowledge of a writer, and some scribes find this task daunt-

---

[16] Agel J (1970) *The Making of Kubrick's 2001*. The New American Library, New York

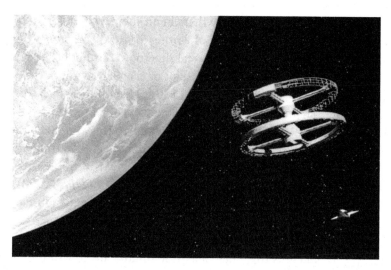

Fig. 3.1 There was no shortage of scientist and engineering consultants on Stanley Kubrick's 1969 film *2001: A Space Odyssey*. Copyright © Metro-Goldwyn-Mayer (MGM). Image credit: MovieStillsDB.com

ing or intimidating. Ashley Miller elaborates, "They take the character, put the character on a pedestal, and don't try to get inside that character's mind. Many screenwriters, don't truly understand what it is to be a scientist, to be an engineer. They don't believe that they have any experience with the analytical process or the analytical mindset. They don't see how it connects with who they are, and, I believe a lot of the time, scientists are written that way because it's biased by the writer's own perception of what a scientist is: somebody who is above, somebody who is an expert, somebody who must be listened to. Maybe they're intimidated by the subject, they think that the subject the scientist character is exploring is, somehow, beyond their ability to understand or convey in a meaningful way, so they get reduced to [spouting] facts."

Writer/producer Kath Lingenfelter says about her experience on *House, M.D.*, "Being a writer, like being an actor, is being a great pretender, I mean you can do as much research as you want, but unless you've lived the life of a lawyer or gone to medical school, when you're writing for a doctor or a lawyer, you're pretending, you're creating an approximation. So it can be very nerve racking in that regard, because you want it to come across as true, you want it to come across as authentic." Screenwriter Bragi Schut adds, "Nobody wants to see the version where the scientists are putting forth ideas that feel like we could come up with them. You want ideas that feel clever and ingenious and intelligent."

It's hard to write people who are smarter than you are.

Kath Lingenfelter, Writer/Producer, *Caprica, House, M.D.*

Zack Stentz adds, "We were first writers in on the new iteration of the *Fantastic Four*. So we were in the position of writing for Reed Richards [whose] real superpower is not that he's stretchy, his real superpower is that he's the smartest guy on the planet. You'll notice there are a lot of clichéd ways of portraying hyperintelligence. It's okay if you don't have a 180 IQ like the character that you're writing, but there are more and less convincing ways of faking it."

One way of not having to fake it, and to not have your main scientist character become an exposition-dumping automaton, is by marginalizing him from the onset, and then killing him as quickly as possible. A singular example of how to marginalize scientist characters is *Red Planet* (2000), about a mission to terraform Mars, one whose outcome may save the Human Race.

The film is openly disdainful of both science and scientists, and scientific missteps are both ubiquitous and frequent[17]. Commander Kate Bowman's[18] opening narration claims that "best scientific minds" are aboard the mission, yet there are only two actual scientists among the crew. Chip Pettingil is a last-minute replacement on the science team (immediately arguing that the best scientific minds are not necessarily on the mission). Chief Scientist Dr. Chantilas' (Terence Stamp) relationship with science is far more troublesome, though. When Gallagher (Val Kilmer) asks:

> GALLAGHER
> You didn't just give up being a
> scientist one day, did you?

Chantilas says that he has all but given up science:

> CHANTILAS
> I realized science couldn't answer
> any of the really interesting
> questions. So, I turned to
> philosophy. I've been searching
> for God ever since. Who knows, I
> may pick up a rock and it'll say
> underneath, "Made by God." The
> universe is full of surprises.

This scene apparently made such an important point that it was repeated as a flashback less than a third of the way into the film.

---

[17] For starters, see "With Mars Movies, at Least They Got the Color Right" in the *Los Angeles Times*, Nov. 19, 2000: http://articles.latimes.com/2000/nov/19/entertainment/ca-54060

[18] The interplanetary missions in both *Red Planet* and *2001: A Space Odyssey* feature a Commander Bowman and an AI run amok.

While nobody would deny Chantilas his spiritual quest[19], one might argue that a mission on which "the hope and survival or Mankind rests"—the "greatest undertaking Man has ever attempted"—does not need a "soul"[20], it needs a contributor, while a Chief Scientist who freely admits to having given up science probably does not qualify. From a mission design standpoint, Chantilas is extra mass that consumes resources and serves no useful purpose. From a screenwriting standpoint Chantilas serves even less purpose than exposition-dumper, he is a philosophy-dumper who gets a few things off his mind, then dies almost immediately after landing on Mars.

The only other scientist on the mission, terraforming expert Chip Pettengil (Simon Baker), contributes no more than Chantilas, he merely lives longer (though scrawny Pettengil does manage to push co-pilot Ted Stanton—a guy built like an NFL halfback—off a towering cliff, thus committing the first murder on Mars). Pettengil has a few lines of dialogue that fall under the category of spacecraft operations, and he perspicaciously observes there is no trace of the algae sent ahead to terraform Mars, even though there should be some hint of evidence. Apart from that, Pettengil had fewer lines of science-related dialogue than any other character in the film, and that includes Lucille the ship's computer. In one scene early in the film, one in which Pettengil might reasonably have offered an insightful or profound science viewpoint, he was passed out drunk.

Clearly *Red Planet* had no need for scientists, they contributed to the body count only. Since Mars in this film had evolved a breathable atmosphere, this military/survival thriller could have easily been the result of a plane crash, and set in the American Southwest. Ironically, despite the film's clear anti-science bias, and attempts to give Chantilas's philosophical musings meaning, what saves the day in the end is, in fact, both science and the human spirit.

Although it is clear from the examples of these two characters that some screenwriters are unsure, or don't care, how to write for scientist characters, whether intentional or incidental, the distain shown scientists in *Red Planet* echoes that which resonanted through the early films of Méliès and Rippert. While there may be nothing to dissuade writers who are overtly anti-science or anti-scientist, the good news is that for writers seeking help with their portrayals of either, advice is far easier to find now than in the past. The scientific community has invested directly in helping writers to depict new types of compelling scientist characters. In 2008, the National Academy of Sciences created the Science and Entertainment Exchange. The Exchange is a science/entertainment matchmaker, connecting Hollywood productions with scien-

---

[19] Just look at Gaius Baltar in *Battlestar Galactica*. The difference, though, is that when Baltar was called on to be a scientist, he always rose to the occasion.

[20] As Chantilas is described by Commander Bowman, played by Carrie-Ann Moss, in the opening narration.

tists of all manners of expertise. Advisors not only counsel producers, directors, and, especially, writers on the technical aspects of their screenplays, they also consult on scientist attitudes, behaviors, and the culture of science. For screenwriters who require assistance with medical concepts and terminology, there is the Norman Lear Center in Beverly Hills, part of the University of Southern Californa's Annenberg School of Communications.

An irony in all of this is that although many screenwriters believe that the life of a scientist is largely alien to them, this really isn't the case. Despite the fact that one discipline falls into the category of what has colloquially been called "left brained" (analytical) versus the other, considered "right-brained" (creative)[21], there are numerous parallels between the two career paths. Both careers generally require an investment of years of study before hopeful practitioners are accepted as peers. Both careers are hypercompetitive. Just as graduate schools weed out only the truly capable and dedicated candidate scientists, there are, according to the Writers' Guild of America West, approximately 3300 people actively employed as screenwriters in the U.S., while, at any given time, there are tens to hundreds of thousands of people trying to "break in"[22].

For a working screenwriter to imagine the life of a scientist, all it takes is a modicum of self-examination and a dash of creativity. For example, as director Jon Amiel shares, "The great problem in the depiction of science is that for every one eureka moment there are 10,000 hours of drudge, and endless looking at computer screens watching data scrolling, and task bars slowly moving across, and dead ends, and disappointments, corrupted data, and all the rest of it which is not essentially [viewed as] the stuff of great drama." For screenwriters struggling with a difficult transition or piece of awkward dialogue, for every eureka moment there are 10,000 hours of drudge, and endless staring at computer screens[23], watching data scrolling[24], and task bars slowly moving across[25], and dead ends, and disappointments[26], and all the rest of it which is not essentially the stuff of great drama (see sidebar "Meet the Storyteller: Ashley Edward Miller").

Stentz observes another similarity between the two careers:

I think it might've been Lawrence Kasdan, who said that being a working writer like having homework every night for the rest of your life. It is a little

---

[21] We know, we know, we did say "colloquially." Deep breath psychologists, we're going to that well again later.

[22] Seriously, everybody in L.A. has a screenplay.

[23] How in the world do I get Fargo into Eureka's Global Dynamics' Sect. 5 in act 4 when Stark revoked his clearance in act 1?

[24] How many pages is this thing up to now? Really? And I still have to write all of act 3.

[25] Because what self-respecting screenwriter would be without Final Draft version 9.1.3.4.2? It's a quantum leap over 9.1.3.4.1.

[26] We're cancelled? But the network said we'd have a full season six!

bit like that, especially being a screenwriter. You're always on the clock. There is always a part of your brain, no matter what else you doing, that's going to be processing on a particular problem that you're working on, or a particular script that you're writing. I think, from talking to scientists, and from knowing a few, that feels very similar to the life of a scientist.

Perhaps it's too much to hope that scientists will prove to be as "interesting" as physicians and attorneys in the near future, but things are certainly looking up for moving scientists in TV and film out of the "functionary" box, and into the realm of fully fleshed out and humanized characters.

Meet The Storyteller: Ashley Edward Miller

How a screenwriter's job is similar to that of a scientist—or, at least, it should be (Fig. 3A.1).

I will start by addressing a myth. The myth is writer's block. It doesn't exist. Writer's block is an excuse. It's something that writers talk about when they don't feel like writing, or they want to pretend that they're not actually unprofessional. Writer's block is something that you lean on when you haven't developed a process to carry you through the day, or the week, or the month, in creating a product. The idea of writer's block, I think, is one that becomes convenient when you don't want the lack of insight to be your fault. It's this magical, mystical thing that you can blame it on, and the reality of it is that being a writer, and comparing it to being a scientist, it's often about simply experimenting. You begin with a hypothesis about your story, about your characters. You explore that hypothesis. You experiment. You see what happens. Sometimes those experiments end in failure. Sometimes they lead you down a road you didn't expect. Sometimes the experiments verify what you believed in the first place.

Fig. 3A.1   Ashley Miller (*left*) and Zack Stentz (*right*) discussing a nuance of writing science fiction screenplays at a 2014 San Diego Comic-Con panel organized by your authors. The hat was Stentz's own idea. Photo: Kristina Grifantini

entist, look at your hypothesis until beads of blood appear on your
y, "Well, I don't really know what to do with this today, because I
lock." It doesn't happen. It doesn't exist.
he matter is that being a professional screenwriter, a professional
ssional writer of any stripe or variety, means developing a process
where you sit down in front of the work, and you are comfortable with creating
hypotheses, you are comfortable with doing the work to prove it or disprove it, you
are comfortable contending with outcomes that are frustrating, and you are willing
to go down avenues that maybe don't fit into the scheme that you originally envisioned. So what both of those activities come down to is discipline, and exactitude,
and professionalism. The difference between a screenwriter and anybody else in
Hollywood is that every single day, a screenwriter must sit down and contend with
that. There is nothing a screenwriter has to wait on. They don't have to wait on
designs, they don't have to wait on a script, they don't really actually have to wait
on notes. They only have to wait on themselves and their own imaginations, and
their process.

So, I would say that being a screenwriter is a lot closer to being a scientist, or being
an engineer, than it is to a lot of other jobs in Hollywood.

3.2 Scientists and Gender/Racial Demographics

The Dudo study cited earlier also found that the scientist characters who appeared in prime-time television were overwhelmingly white males. In the
United States in 2001, 82 percent of scientists were white, 2.5 percent were
black, 13 percent Asian, and the remaining 3 percent were Hispanic or Indian. Within those classifications, 72 percent of scientists were men and 28 percent women. Ten years later, in 2011, little changed: 81 percent were white,
2.75 percent were black, 12.5 percent were Asian, and the remaining 4 percent were Hispanic, Indian/Alaskan Native, or Other[27]. Gender representation, while better for women, was not significantly better: 68 percent were
men; 32 percent were women. These statistics suggest an important question:
does Art simply imitate life, or, like the science accuracy issues we examined in
Chapter Two, might there exist another type of feedback loop that influences
people's career choices and maintains the status quo?

Let's start off answering this question by looking at a character who has
been depicted numerous times, in many works of television and film: Captain
Nemo from Jules Verne's novel *20,000 Leagues Under the Sea*, first published
in 1870. Although it is arguable whether Nemo is a scientist, or just very
science-literate, the character is often cast as Caucasian, while in the novel,
Nemo was the Indian Prince Dakkar, son of the Raja of the Kingdom of Bun-
delkund—a region in central India. To date the only Indian actor to play him

---

[27] Racial categories were those reported by the National Science Foundation, as was the change in the
categories reported between the two surveys.

on screen has been Naseeruddin Shah in *The League of Extraordinary Gentlemen* (2003). More recently, although stuck in development hell for years, attempts to make a live action version of the 1988 Japanese animated classic *Akira* (set in a future version of Tokyo) have generally had Caucasian actors attached to the cast, while the two most recent *Godzilla* remakes (in 1998 and 2014) have had Caucasian leads. *Pacific Rim* (2013) deserves credit for at least having an East-Asian co-lead (and an unusually well-developed female character as well) in Rinko Kikuchi's portrayal of Mako Mori.

The practice of taking characters of color and casting them as Caucasian in screen adaptations is not unique to science fiction of course, and it happens enough to have earned a specific term: *whitewashing*. Another problematic area is the portrayal of the gender of scientists, especially in media targeted at adolescents and adults. Children's shows that depict scientists nearly always show men and women as equals in all respects. Not so in prime-time shows and blockbuster films.

Lynda Obst was the producer for the 1997 film *Contact*[28]—one of the better films ever in its depiction of science and scientists—says that she believes the film would have gotten off the gound far easier had Ellie Arroway been *Mel* Arroway (Fig. 3.2):

> One of the things that took it so long is that I had a *female* scientist. At the time, there weren't that many women who could open a movie. There were

Fig. 3.2   Jodie Foster is Dr. Ellie Arroway in *Contact* (1997). Copyright © Warner Brothers Pictures. Image credit: MovieStillsDB.com

---

[28] As well as the 2014 film *Interstellar*.

many more men that could open a movie, and the fact that my lead was female meant that only one or two or three women could play the part. If my lead had been male, it probably would have been easier to get the movie made, I think people felt, well, maybe guys wouldn't come to a male-dominated genre if a woman was in the lead. Fortunately, there was an actress named Jodi Foster, who was the perfect Ellie Arroway, and she knew that immediately. She could open a movie, and so we could get it made.

Although the portrayal of chaotician Dr. Ian Malcolm in the first two *Jurassic Park* movies is contentious—some viewers loved the character, others found his quirks annoying[29]—there are worse things for a scientist character to be than a quirky exposition machine: enter Julianne Moore's character, Dr. Sarah Harding. In *The Lost World: Jurassic Park* (1997), while Dr. Harding was ostensibly a hotshot behavioral paleontologist, the character more accurately served as a slightly-more-competent-than-usual damsel in distress. In fact, it was Dr. Harding's (supposed) intelligence that put her in distress in the first place. Bring the baby T-Rex aboard the mobile home? Sure! Mommy the eight-ton predator will never smell her baby or hear its cries there! When Dr. Harding's scientific short-sightedness leads to trouble, she then proves that she can wail on par with any B-movie scream queen[30].

> I mean c'mon, what girl didn't want to be Dr. Dana Sculley?
>
> Kath Lingenfelter, Writer/Producer, *Caprica, House, M.D.*

Jaw dropping as it may be, an even less flattering portrayal/caricature was Nuclear Physicist Dr. Christmas Jones, played by Denise Richards. Zack Stentz shares, "She's a lovely person and fine actress, but spent half of *The World is Not Enough* doing science in bicycle shorts. I think that particular piece of casting has come to be shorthand for not very credible female scientist—female scientist cast for hotness more than credibility as a scientist." Yvonne Strahovski similarly suffers as being cast as a not-so-mad successor to Doctor Frankenstein in 2014's *I, Frankstein*, where she plays a somewhat thinly written romantic interest for the reanimated lead character. This is especially ironic given Strahovski's performance as Agent Sarah Walker in the spy-fi television series *Chuck* (2007–2012), where she demonstrated that she can deliver on playing a technically skilled character who is no-one's damsel in distress.

---

[29] On the topic of Dr. Malcolm. Author KRG: thumbs up; Author SAC: thumbs up. It's official: Malcolm rocks.
[30] Dr. Malcolm did predict that later there would be "… running and screaming."

However, there are some positive signs in recent years. In the Mar
books, Jane Foster is a nurse, but in the 2011 superhero film *Thor*, the character
acter is an astrophysicist. When asked if Marvel resisted the change, *Thor*
co-writer Stentz, says, "[Marvel] were fully on board with [switching Foster's
career]. In talking to the people at Marvel, everyone pretty much thought that
physics would be a great form of connection between the weirdness of Asgard
and that whole universe, and our universe—or at least the Marvel version
of our universe. It made more sense to have Jane Foster be more of an equal
partner, and if you made her a physicist then you could have Thor intersect
with physics problems that she's trying to solve. If anything, the only problem
that it caused was that it became really imperative to get an actress who could
credibly play that part." Stentz shared that director Kenneth Branagh was
also onboard with the career switch, but adamant, "Well the one thing that
we can't have is 'Denise Richards: Nuclear Physicist.' So we were delighted
when we found out they had gotten Natalie Portman who is quite intelligent,
poised, and Ivy League educated." (Portman, whose given name is Hershlag,
has two published scientific papers, one first-authored, to her name[31,32].)

Although their numbers are increasing—e.g. Kathryn Janeway (*Star Trek:
Voyager* 1995–2001), Samantha Carter (*Stargate: SG-1*, 1997–2007) Holly
Marten (*Eureka* 2006–2012), Amy Farrah-Fowler and Bernadette Rosten-
kowski (*Big Bang Theory* 2007–), Julia Walker (*Helix*, 2014–2015), etc.—and
their portrayals are typically of competent characters, male scientists still far
outnumber female scientists. In Chapter Two, we saw how the Hollywood
Curriculum Cycle helps to propagate inaccurate science tropes repeatedly, and
it is clear how this similarly leads to reinforcing racial and gender equilibria as
well. Psychologist Andrea Letamendi shares, "We gain a lot about our identity, and our value to society, based on what's portrayed in the media." Kaplan
elaborates:

We know from research that when people watch movies or television shows,
it does affect what they know, what their attitudes are, and how they behave.
They get, through depictions, a version of reality, which, even if they tell themselves they know is made up, it still affects them. So that the depictions create a
universe which the viewers think is accurate. When messages appear, messages
which purport to be about, oh, the symptoms of a disease or a fact in history—
even though they know they are dealing with something that was written by
screenwriters, and could be entirely fictional—there is the tendency, which has

[31] Hershlag N, Hurley I, Woodward J (1998) A Simple Method to Demonstrate the Enzymatic Production of Hydrogen from Sugar. *Journal of Chemical Education* 75(10): 1270
[32] Baird AA, Kagan J, Gaudette T, Walz K, Hershlag N, Boas DA (2002) Frontal lobe activation during
object permanence: data from near-infrared spectroscopy. *NeuroImage* 16(4): 1120–6

been measured, to think that what they hear is accurate, and it affects their view of what is true.

There have been few research studies that have examined how media depictions of scientist characters influence public attitudes, but one that did was a 2009 study by Losh[33]. What that study found was that as scientist characters have evolved from move-negative portrayals since the 1990s to the more positive portrayals of today, stereotypes held by adult audiences have evolved similarly. The study found that, compared to two decades prior, American adults in the 2000s were far more likely to feel that careers in science were desirable for both their children and for themselves.

While there exists a fair number of programs to encourage minority and women to consider STEM (Science, Technical, Engineering, Mathematics) careers, working scientists frequently cite works of Hollywood science fiction as prime motivators for their educational and career choices. If our culture places value on encouraging women and minorities to pursue careers in STEM fields, then Hollywood can help, says Kaplan:

I spoke at a conference in Berlin whose purpose was how to get women into STEM careers, and one piece of the agenda was, "How about if we depict women *in* STEM careers in entertainment?" There is a field called entertainment education. It is best known and practiced outside the US, The US is the outlier, because here our entertainment industry—because of the First Amendment and other reasons—is a for-profit, fiercely independent, artistically driven, enterprise. In other countries, there are state-owned TV stations, state-owned movie companies, state-owned studios, and subsidized product. In many countries around the world, there are very careful plans to use their entertainment in order to accomplish goals—whether the goal is adult literacy, or stopping spousal abuse, or HIV testing. In those countries, they mount campaigns using [television] shows in order to do it.

In a similar vein to "If you build it, they will come," one might say "If you show it, they will believe." More woman and minority students may consider science as a career if Hollywood simply serves up more Ellie Arroways[34], Miles Dysons[35], Greg Pratts[36], Kim Andersons[37], Richard Daystroms[38], and even vil-

[33] Losh SC (2012) Stereotypes about Scientists over Time Among US Adults: 1983 and 2001. *Public Understanding of Science* 19(3): 372–382
[34] Jodi Foster's character in *Contact (1997)*
[35] *Terminator 2: Judgement Day (1991)*
[36] Mekhi Phifer in *ER*.
[37] Tamilyn Tomita in *Eureka*
[38] *Star Trek*: The Original Series episode "The Ultimate Computer".

lainous—or at least morally ambiguous—Captain Nemos, Hiroshi Ha Meh Yewlls[40], and Henry Deacons[41,42].

> The unreal is more powerful than the real. Because nothing is as perfect as you can imagine it. Because its only intangible ideas, concepts, beliefs, fantasies that last. Stone crumbles. Wood rots. People, well, they die. But things as fragile as a thought, a dream, a legend, they can go on and on. If you can change the way people think. The way they see themselves. The way they see the world. You can change the way people live their lives. That's the only lasting thing you can create.
>
> Chuck Palahniuk, Author, *Choke*

Depicting more female scientist characters is a start in the right direction, but it is, as a mathematician would say, a "necessary but insufficient" condition, and will not guarantee that university registrar offices will be flooded with thousands of young women yearning to pursue careers as physicists or geologists or astronomers. Another important component to this is that the depiction needs to be an attractive one as well. The same 2009 study by Losh found that the way many scientist characters are portrayed can be off-putting to women: "Although scientists are typecast as clever or diligent, and surveys indicate that science careers are prestigious, scientists are also described as socially inept or even dangerous. Images of scientists not only portray them as male, but also often depict science careers as uninviting." Losh adds, "Media… may depict lonely science careers that can seem particularly daunting for young women." Serving up women scientist characters in strong, competent, socially connected depictions—two excellent recent examples being *Eureka*'s Dr. Allison Blake (Salli Richardson) and *Stargate*'s Dr. Samantha Carter (Amanda Tapping)—is useful in evolving the question "Do I believe I *could* become a scientist?" into "Do I *want* to become a scientist?" in the minds of STEM-inclined young women.

The main character of the film *Contact*, Dr. Ellie Arroway, was depicted as being incredibly dedicated to her work, her career, and to discovering fundamental truths about the Universe, yet it was also clear in that film that she did not lack for social contact with friends and peers, making science appear

---

[39] *Helix* (2013–)
[40] On *Defiance*, Doc Yewll, played by the incredible Trenna Keating, is actually an alien—an Indogene. We're just curious how closely you're paying attention.
[41] During his evil period on *Eureka*
[42] *Eureka*'s Henry Deacon and *Terminator*'s Miles Dyson were both played by the very talented Joe Morton.

like an appealing career option in that film. Of *Contact*'s legacy, producer Lynda Obst says, "One of the things that make me happiest about *Contact* [is that] there are very few movies that have an actual impact in the world, beyond their opening first two weekends. It's just amazing that there are so many more women in astronomy, and in the sciences in general, than there were when we started to talk to women scientists about what it was like when they were first in graduate school, and there were no other women in the class. Now I count women astrophysicists and cosmologists and physicists as some of my best friends."

> It's not great lives that make great movies, it's a great connection between the filmmaker and the subject that makes great movies.
>
> Jon Amiel, Director, *Creation*

3.3 Hollywood Scientists as Idiosyncratic Nerds

nerd\nərd\noun: an unstylish, unattractive, or socially inept person; *especially*: one slavishly devoted to intellectual or academic pursuits <computer *nerds*>[43]

Nerd (adjective: nerdy) is a descriptive term, often used pejoratively, indicating that a person is overly intellectual, obsessive, or socially impaired. They may spend inordinate amounts of time on unpopular, obscure, or non-mainstream activities, which are generally either highly technical or relating to topics of fiction or fantasy, to the exclusion of more mainstream activities.[44]

The first documented appearance of the word *nerd* is the name of a creature in Dr. Seuss's (Theodor Geisel)[45] 1950 book *If I Ran the Zoo*. The slang meaning of the term dates back to 1951, when *Newsweek* magazine reported on its widespread use in Detroit as a synonym for somebody who wasn't "with it,"— a "square"[46]. Television popularized the term, in the 1970s, as frequent use in the sitcom *Happy Days*[47] led to more widespread acceptance[48].

---

[43] Merriam-Webster Dictionary
[44] Wikipedia
[45] Geisel, Theodor Seuss (1950) *If I Ran the Zoo*. Random House Books for Young Readers, New York, p 47
[46] *Newsweek* "Jelly Tot, Square Bear-Man!" (1951-10-8), p 28
[47] Fantle D, Johnson T (2003) "Nerd" is the Word: Henry Winkler, August 1981, in: *Reel to Real: 25 Years of Celebrity Interviews*. Badger Books, pp 239–242
[48] Among those not so-labeled, that is.

> And then, just to show them, I'll sail to Ka-Troo. And Bring Back an It-kutch, a Preep and a Proo, a Nerkle, a Nerd, and a Seersucker, too!
>
> Theodor Geisel (Dr. Seuss), *If I Ran the Zoo*

"Nerd" is a stereotypical term, for a person who is intellectual, often technically inclined, and often socially awkward (In part because it had such as large population of nerd characters, *Freaks and Geeks* (1999–2000) portrayed one nerd character, Bill Haverchuck (Martin Starr), as socially awkward without any notable intellectual or technical gifts). It differs from the British "boffin"—originally slang for scientists and engineers in World War II—in that boffins are the fine fellows who did the technical and scientific work that won the war and saved the day; "Nerd" is a derogatory term for somebody who, truth be known, has many of the same personality traits as a boffin, without the heroic overtones. As with so many similarly pejorative expressions through history, the target group has embraced the term, and now wears the nerd moniker as a source of pride, group identity, and fellowship. One could easily envision a softball team at JPL, a bowling team from Los Alamos National Laboratory, even a Rube Goldberg Machine team from Purdue, called the "Idiosyncratic Nerds."

When "computer nerds" became fantastically wealthy during both the PC boom of the 1980s and 1990s, and then again during the dot-com boom of the 2000s, suddenly nerds became cool. It an odd role reversal, now people who've played *Halo* once, read an *X-Men* comic once, or enjoyed the *Transformers* movies—many of whom probably picked on the Poindexters in high school—proudly boast, "Dude, I'm, like, a total nerd."

> So you're a little weird? Work it! A little different? Own it! Better to be a nerd than one of the herd.
>
> Mandy Hale, Author

When James Cameron lamented that scientists are, too frequently, portrayed as "idiosyncratic nerds," clearly he referred to the less flattering characteristics of nerdlihood, in a time when "Nerd" or "Geek" was almost exclusively pejorative. Assuming Mr. Cameron is right[49], why where scientists portrayed in that manner in Hollywood productions?

---

[49] Let's face it, Mr. Cameron has some pretty serious Nerd Cred in his own right.

From the early days of film until the early 1990s, scientist characters, when they were not evil or megalomaniacal, existed largely to deliver exposition, they were relegated to being explainers. When one understands how screenwriters are trained, the situation becomes ever worse for our scientist characters, at least as far as them actually acting like scientists. Television writer, and *Star Trek* science advisor, Andre Bormanis explains, "Exposition is the term for dialogue that is explaining to the audience what is happening at that stage of the story. Exposition, in the long history of drama, has been considered the thing you most want to minimize. 'The ideal script has zero exposition.' This is what every screenwriting professor, for time immemorial, will tell you—that exposition is deadly, because it brings the action to a stop, and the minute you do that, the audience starts to tune out." Lingenfelter adds, "As soon as you stop to give a science lesson that seems a little overt, people are going to run screaming from their television." Bormanis continues, "So when you do exposition, you have to find clever ways to do it. You've got to kind of disguise it a little bit: do it in a scene that's very compelling in, and of, itself, or that has some humor in it, or that has some other kind of tension going on in the background that's clear to the audience so that one or two lines of exposition is not going to release that tension, then you can get away with it."

So today young screenwriters are counseled to hide exposition as much as possible. Put it in the background, show it as a graphic or, better, as a visual effect, work it in as part of the character drama. In years past, however, exposition was delivered more obviously, and it fell to scientist characters to explain what was happening—or what might happen—and this was often front-loaded at the beginning of the narrative. The rest of the time there was little for them to do, except on rare occasion when your scientist characters would step from the shadows around the end of the second act, deliver a line or two of exposition to push the plot forward, and step back. Because the writer couldn't have them standing around with their hands in their pockets, literally or figuratively, they were marginalized by being quirky, or the parent of the lovely female romantic lead, or killed off. Marty Kaplan elaborates:

> If you're a screenwriter, you first have the job of making every single one of your characters memorable, and that almost always involves some kinds of quirks, ticks, and eccentricities. So, scientists, welcome to the party. That's how writers work. You've got all these different people, and your audience has to be able to keep track of them, and you have to seem like you're not moving cut out paper dolls. The way you do it, is to every character you assign something that separates them from the crowd. If the scientist is a lead in the story then that character is going to have a past, and a family, and a conflict, and a set of issues, and be more full-blooded. If that character is just somebody who moves

through, then that character will get the same short shrift of other characters that move through: the explainers, the reporters, the magical black person, the sassy waiter, the knowledgeable cab driver, these are all tropes. So scientists are susceptible to that same kind of treatment.[50]

Today, while scientists may still be explainers, they tend to be, like the rest of any ensemble or movie cast, more complex in their characterizations. Ashley Miller says, "Now, in some cases, for example Walter Bishop, you allow that character to serve that function [explainer], and have that authority as a scientist, and then you build in character issues to get the audience behind them emotionally in a different way." So although the depiction of scientist characters has clearly evolved, it is also clear that, even today, their depictions are often unambiguously "idiosyncratic." The difference between then and now is that, in Hollywood's earlier days, scientists were made to be quirky because writers knew little else to do with them, while today it is more of an informed creative decision with the door left open for deeper character development.

Clearly "idiosyncratic" applies to scientist characters, but are they depicted as nerds? Cutting to the core of the definitions that kicked off this section: Are scientist characters "…slavishly devoted to intellectual or academic pursuits, to the exclusion of more mainstream activities?" Definitely. What about real-world scientists? Definitely—either by choice or necessity, it *is* a highly competitive career path. "Particularly of a technical nature?" By definition, for both. Do they spend inordinate amounts of time on "…obscure, non-mainstream activities?" Yes. Do real world scientists? Absolutely, and most wouldn't have it any other way.

> Whenever I think of how much pleasure I have interviewing scientists, I remember that they're having the real fun in actually being able to do the science.
>
> Alan Alda, Actor/Director

In an August 1999 article in the *New York Times* that profiled planetary scientist and Team Leader for the Cassini Imaging Team, Carolyn Porco[51], she said, "There are absolutely no high-maintenance items in my house of any

---

[50] This actually still happens from time to time. A prime recent example is the South Korean time travel film *AM 11:00*, where the exposition is obvious, front-loaded, inaccurate, and self-contradictory. While, to the film's credit, the scientist characters are presented as real, flawed, people, if you have a choice between watching this film and a rerun of "Spock's Brain," bolt for "Brain and brain, what is brain?" We've suffered so you don't have to. You're welcome.

[51] Who was also the science advisor for *Star Trek* (2009) and *Star Trek: Into Darkness* (2013).

kind," she added, "plants, pets or husbands." Porco subsequently shared that because of the nature of the job she had to, "Clear the decks of any semblance of a normal life to get the job done, and done well." While some scientists and engineers happily juggle outside interests and relationships with their work, this kind of solitary dedication to their field is far from unknown. (It's also not unknown among high-achievers in other fields, such as painting, athletics, finance, ballet, politics and so on.)

In *I, Origins* (2014), written and directed by Mike Cahill, a fair bit of effort is made to portray scientists as real people; something Cahill says was influenced by the fact that he has two brothers who are scientists[52]. *I, Origins'* scientists go to parties, make questionable personal decisions, and so on. Speaking at a screening of the movie, Cahill said: "I wanted to show them at a party because scientists are real people…in the first ten minutes I wanted to show scientists as scientists getting laid … to show a real Ph.D. student." Nonetheless, when the same Ph.D student interrupts an important romantic moment with his girlfriend to race to the lab because his lab partner has called with exciting research news, it's also entirely in character.

> I have a girlfriend. Her name is science.
>
> Dr. Plotkin, *Eureka,* "One Giant Leap"

The scientists in the film *Twister* (1996) repeatedly, and almost obsessively, risk their lives to introduce sensors into the funnel of a tornado—in the hopes that measurements of internal structure of a tornado will lead to a prediction system that will, ultimately, save lives. Although main character Bill Harding is portrayed as the "cool" atmospheric scientist who got out of research to take a job as an on-air meteorologist, when research results are on the line, he is also shown as the most dedicated, technically knowledgeable and, at some level, "nerdy" of his quirky band of scientists and technicians[53] (Fig. 3.3).

There are two other noteworthy points in *Twister*. When a former colleague leaves the group, and obtains corporate sponsorship to perform the exact same kind of research, one of our protagonists quips, "He's in it for the money, not the science."[54] The other significant aspect of this film is one that

---

[52] *I, Origins* won the 2014 Sundance Film Festival's Alfred P. Sloan Prize, awarded to an "…outstanding feature film that focuses on science or technology as a theme, or depicts a scientist, engineer, or mathematician as a major character."
[53] One might expect that the scientist characterizations in this film were better than usual, the screenplay was co-written by Michael Crichton.
[54] HA HA HA HA HA! HA! HA HA! As if.

Fig. 3.3 Helen Hunt and Bill Paxton as scientists Dr. Jo Harding and Bill Harding in *Twister* (1996). Copyright © Warner Brothers Pictures. Image credit: MovieStillsDB.com

reflects an aspect of the lives of real scientists—and which is also displayed prominently by the scientists in the movie *Contact*, the grad students in *Real Genius*, and even the characters in *The Big Bang Theory*—is that although scientists may appear to be socially awkward, it often appears that way only to people who don't speak their language. Among other scientists and people who speak "Tech," social interactions come easily, and they are often socially well-connected. Social skills actually count for a lot in a science career—modern science is a highly collaborative affair, where large, even international teams are becoming increasingly commonplace. Scientists who don't work well with others generally do not get to play with million- and billion-dollar instruments and other toys.

James Cameron did, in fact, make very similar observations about real scientists in the same 2005 interview where he referred to Hollywood scientists as "nerds and villains," when he noted, "One of the things we tried to do with this film [*Aliens of the Deep*] was to show what scientists are really like, which is they're people. And sometimes they're really cool people, and they really care about what they're doing. They're not driven by a materialistic value system. They're seeking something else, something more important. That message speaks to me."

What exactly that "something else" is, may not necessarily be a saint-like desire to increase human knowledge selflessly. One of author SAC's professors[55] once explained to a television audience that scientists were motivated

---

[55] J.M.D. Coey, who is also the lead author on SAC's single published scientific paper before the cheap thrills of journalism lured him away.

by two personal drives: curiosity and ambition. In other words, every scientist wants to be the one who figures out the answer to some question about the universe that they find personally fascinating.

Two other screen characters that would certainly be considered in the social mainstream are scientist/industrialists Daniel Graystone[56] (Eric Stoltz) from *Caprica* and Tony Stark (Robert Downey, Jr.) from the *Iron Man* series of movies. These characters are alike in many fundamental ways. Both are portrayed as fabulously wealthy, living in magnificent mansions. Both run their own companies. Both have cutting-edge laboratories within their homes. Both are portrayed as media-savvy. Both are shown interacting readily with people of all levels of technical know-how.

> People think of the inventor as a screwball, but no one ever asks inventor what he thinks of other people.
>
> Charles Kettering, Engineer/Inventor.

Although both are insanely rich, both Graystone and Stark are depicted as possessing the same type of work ethic and single-minded dedication as our heroes from *Twister*. In both cases, either explicitly or implied, the viewer knows that when there's hard work to be done, or a particularly challenging problem to be solved, there's nobody more willing to loosen his tie, roll up his sleeves, and put in whatever long hours are necessary to attack it.

At the opposing end of the "Nerd Spectrum" is *CSI: Crime Scene Investigation's* Dr. Gil Grissom. The character of Gil Grissom often does not recognize societal norms of behavior; he has little interest in mainstream media, generally does not participate in office politics, and often shows distain, or at least little interest, in human emotional connection. What this character shares with Stark and Graystone is an intense devotion to his career and duties (Fig. 3.4).

Which is the more accurate depiction of an actual scientist? Perhaps a more insightful question might be: what is the fundamental difference between a socially nimble Dr. Graystone and the less-socially adroit Dr. Grissom? The simple answer might be, surprisingly, "choice." Zack Stentz shares, "I talked to [William Petersen, who plays Grissom] in a party about whether his *CSI* character was on the [autism] spectrum, and he was very adamant that he was not, that he just viewed [his portrayal] as being a choice by character who is prioritizing other things." A recent study by researchers from the University of Illinois, the Universidad Autónoma de Madrid, and the Rehabilitation Insti-

---

[56] KRG: You have to love any series where the main character is a tall, red-headed, computer scientist.

Fig. 3.4 Esai Morales and Eric Stolz are Josef Adama and Dr. Daniel Graystone in *Caprica* (2009). Copyright © David Eick Productions and Universal Media Studios. Image credit: MovieStillsDB.com

tute of Chicago suggests that William Peterson's creative choices in portraying Gil Grissom may have been more astute than he imagined.

Assume that a de facto job requirement is that scientists have high cognitive intelligence, meaning that they learn quickly and perform well at problem solving, making future plans, abstract thinking, and comprehending abstract concepts. Until recently neuroscientists have considered cognitive intelligence and emotional intelligence—the ability to recognize and cope with emotions within the self and those displayed by others—as independent. The Illinois study found that cognitive intelligence and emotional intelligence are actually strongly correlated: as one rises so does the other. The study's[57] lead author, neuroscientist Aron Barbey from the University of Illinois, explained, "Intelligence, to a large extent, does depend on basic cognitive abilities, like attention and perception and memory and language … But it also depends on interacting with other people. We're fundamentally social beings and our understanding not only involves basic cognitive abilities but also involves productively applying those abilities to social situations so that we can navigate the social world and understand others."[58] In other words, scientists have the tools to be at least as socially adroit as anybody else, but, in many instances, simply choose not to be. The characters of Stark and Graystone made one choice; Grissom made another.

---

[57] Barbey AK, Colom R, Grafman J (2012) Distributed neural system for emotional intelligence revealed by lesion mapping. *Soc Cogn Affect Neurosci*. doi: 10.1093/scan/nss124
[58] *Good News, Nerds! IQ Linked to Emotional Intelligence*: http://www.medicaldaily.com/good-news-nerds-iq-linked-emotional-intelligence–244372

> Nerd: One whose unbridled passion for something defines who they are as a person, without fear of other people's judgment.
>
> Zachary Levi, Actor/Director

The question though remains, "Are scientist characters often depicted as nerds?" The answer is, "Probably, but does it matter?" There certainly is enough of an overlap in the definition of "nerd" and the behavior of scientists—both real and on screen—to understand how those lines may get blurred in the shorthand that is a screenplay or teleplay. At the same time, the complaint that Hollywood depicts scientists as "idiosyncratic nerds" has, in many important respects, been rendered inert by time and circumstance. Even by the time James Cameron made his interview quip, Hollywood was well into evolving past tired scientist stereotypes, and has, in fact, graduated from mocking nerds to celebrating them[59]. Given society's increasing reliance on high technology, this, probably neither a trend nor a fad. Nerds are here to stay, and—as long as being a nerd is decoupled from the stereotype of being a badly dressed social loser, or exclusively associated with being male and white—that may not be so bad.

> Yes, I am a nerd.
>
> Gale Boetticher, *Breaking Bad*, "Sunset"

3.4 Hollywood Scientists as Active Villains

> Gee, Brain, what do you want to do tonight?
> The same thing we do every night, Pinky, try to take over the world!
>
> Pinky and The Brain, *Pinky and The Brain*

From the first depiction of Dr. Frankenstein in 1910 and Rotwang in *Metropolis* (1927), until the early 1990s, insane, megalomaniacal, misguided, or power-hungry scientists were fixtures in science fiction and horror films. The term "mad scientist" conjures images of a crazy-haired, wild-eyed, bespec-

---

[59] We acknowledge that there is a debate, even between your authors, about which of these happens on *The Big Bang Theory*.

tacled scientist in a lab coat, scheming to take over the world, cheat death, or alter the nature of what it means to be human, while lurking in his dungeon laboratory surrounded by Van de Graff generators, beakers of boiling chemicals, complicated gadgets, and/or jars with unidentifiable (or, worse, identifiable) body parts.

Many non-scientists find scientists scary, or at least intimidating. Scientists work at the arcane boundary of what's known. They push back the boundaries of human knowledge. When a scientist makes a discovery, that scientist is, for a short while, the only person in human existence with some piece of knowledge or insight into a universal truth. They have access to cutting edge technology. Scientists have made the breakthroughs that enable the creation of some of the most fearful weapons known to humanity.

These things make people uncomfortable, and where people are uncomfortable, there is a wellspring for drama. Would *Helix*'s Dr. Hiroshi Hatake seem as menacing were he a 500-year-old professor of English Literature? Would *Doctor Horrible's Sing-Along Blog* (2008) have been anywhere near as popular had it been *Mister Horrible, Esq.'s Sing-Along Blog*?[60] Does a plumber credibly have the means to create a device that might allow him to take over the world? Assuming he did, is the resulting filthfest even a story we want to watch?[61] Albert Einstein once said, "Those who have the privilege to know, have the duty to act." Isn't it just a hop, step, and small quantum leap from that to, "The world is a mess, and I just need to *rule* it."?

The Dudo content analysis study we cited previously found that scientists in TV today are frequently cast in good or mixed roles rather than as evil scientists. That has, certainly not always been the case. The idea that there exists "forbidden" knowledge, or knowledge that is too dangerous for mere mortals to possess, is a theme that resonates back thousands of years and personified in the forms of Prometheus, Daedalus, Adam and Eve, Faust, and others. Usually the lure of possessing such knowledge is seductive, so characters will sacrifice morals and integrity in its pursuit, even though to seek or possess such knowledge invites punishment, death, or damnation. In cinematic history, it only made sense that the tragic character in pursuit of secret or forbidden knowledge would be a deranged scientist.

There is a recognized phenomenon called *scientist distrust*[62] that goes a long way towards explaining Hollywood's long-time fascination with the mad scientist. After World War I, which saw the introduction of the airplane and

---

[60] Ignoring, for the moment, that vast legions of fans would watch a dramatic reading of the phone book if it was produced by Joss Whedon.

[61] Then, again, there was the *Toxic Avenger* (1984).

[62] It turns out that Americans distrust science journalists even less. Would we lie to you?: http://www. huffingtonpost.com/2013/12/21/faith-in-scientists_n_4481487.html

poison gas into warfare, the notion that a single person, in particular a scientist, could bring about the end of the world became entrenched in humanity's collective psyche.

A 2005 survey revealed that, in 1000 horror movies distributed in Britain from 1931 to 1984, mad scientists or their creations were the villains or monsters in 31 percent of them (41 percent if mutations are factored in); in 39 percent of the movies, the menace was a result of scientific research. Scientists were heroes in only 11 percent[63] of those movies and, in most of those cases, the movies were pre-1960. By the 1960s, the zeitgeist was infused with the memory of the sadistic experimentation of Josef Mengele[64] during the Nazi regime, alongside Cold War nuclear warfare fears. The evil scientist character became firmly entrenched in cinema.

> Do you know, it's amazing how many supervillains have advanced degrees. Graduate schools should probably do a better job of screening those people out.
>
> Dr. Sheldon Cooper, *The Big Bang Theory*, "The Codpiece Topology"

Although the general public has a short collective memory, and sometimes an even shorter attention span, they've not forgotten that, at the core of nuclear weapons—and the equally scary chemical weapons and bioweapons—lay the foundations of scientific inquiry. In more recent times, large segments of the U.S. population are openly skeptical on science-related topics such as the existence and cause(s) of global warming, the health risks of vaccinations, and the dangers of genetically modified organisms[65]. Fairly consistently, the "flavor" of distrust is highly correlated with party affiliation. On the topic of genetically modified organisms[66] and food, a major complaint is that science does not understand the potential side effects of gene modification, science is venturing into a dangerous territory, science playing with knowledge that should best be left alone. Not surprisingly, the prototypical mad scientist was Dr. Frankenstein, and opponents of GMO-based crops call it "Frankenfood."

---

[63] Frayling C (2005) Hollywood's changing take on the scientist. *New Scientist*, p 48

[64] There were other real world mad scientists like Dr. Vladimir Demikhov, Dr. Sergei Brukhonenko, and Dr. Shiro Ishii.

[65] In the United States, the political right tends to produce the most evolution and climate change deniers, while the political left produces more anti-vaxxers and anti-GMO activists. So, depending upon the topic, both sides are either equally guilty of anti-science rhetoric, or equally willing to accept the view of an overwhelming majority of Ph.D.-holding scientists.

[66] Did you know that there is no such thing as a wild Guinea Pig? They are GMOs, bred to be a source of food, dating back to about 5000 BC.

The evil scientist still makes the occasional appearance in TV and film and, some would claim, in the grocery store aisles as well.

> There's a fine line between genius and crazy.
>
> Jack Carter, *Eureka*, "Bad to the Drone"

3.5 Modern Hollywood Scientist Archetypes

In a 2008 symposium lunch talk to a combined group of scientists and entertainment industry professionals in Century City, CA, a famous celebrity scientist began, "I was having dinner a few nights ago with a friend who's an entomologist. Then I thought, 'Entomologists. Why don't you ever see a TV show that stars an entomologist?'" While most of the audience laughed, one smartass in the crowd[67] yelled, "*CSI*." Time out. A bug[68] doctor is the star of a mainstream network TV series? It's true. Although the speaker was unfamiliar with the lone counterexample to his quip[69], that such a counterexample[70] exists is a clear demonstration of the huge strides scientists have made in their Hollywood portrayals in recent years.

Although an observer can always cherry pick and find examples in both TV and film where scientist portrayals have regressed to time-worn stereotypes, media analysts Nisbet and Dudo[71] state, "Depictions prior to the 1990s featured some of the most negative archetypes, yet over the past two decades, the most positive archetype—scientist as hero—appears with increasing frequency as a central character both in film and television. This trend suggests that somewhat contrary to scientists' impressions, their image is not as negatively slanted as they might presume.

We previously mentioned that a content analysis study by Dudo and colleagues[72] found that scientist characters are rare in prime-time television to-

---

[67] Who may have been one of your authors.

[68] Yes we're aware that all bugs are insects, but not all insects are bugs. We're going with the colloquial usage.

[69] Scientists are very dedicated to their careers, remember? Most watch little television.

[70] There was also the lovely Dr. Bambi Berenbaum (Bobbie Phillips) in the *X-Files* episode, "War of the Coprophages", but that casting was on par with Denise Richards: Nuclear Physicist. Even Ms. Phillips's IMDb profile describes the character as, "…she was the unforgettable Dr. Bambi Berenbaum we all loved to hate"

[71] Nisbet MC, Dudo A (2013) Entertainment Industry Portrayals and Their Effects on the Public Understanding of Science, *Hollywood Chemistry: When Science Met Entertainment*, pp 241–249.

[72] Dudo A, Brossard D, Shanahan J, Scheufele DA, Morgan M, Signorielli N (2011) Science on Television in the 21st Century: Recent Trends in Portrayals and Their Contributions to Public Attitudes Toward Science, *Communication Research* 48(6): 754–777

me study also found that, when they are depicted, it is nearly al-
:ive light. Of the scientist characters these researchers observed,
re characterized as good, 26 percent as both good and bad, and
just 3 percent as bad.

Science fiction novelist David Brin, who also holds a Ph.D. in physics from
UCSD agrees: "In the context of Hollywood's reputation, as a producer of
lurid, lowest common denominator entertainment, I would have to say that
science has very little to complain about, at least as scientists themselves are
portrayed. They are more often viewed as heroic figures, seeking knowledge,
issuing warnings that aren't heeded by the political or corporate caste. So the
opportunity for scientists to see someone in their profession portrayed in a
positive light onscreen … this happens a lot."

Researchers who study media identify six[73] types of scientist archetypes in
TV/film today. Occasionally screenwriters blur the lines between these archetypes—incorporating elements of one or more into a single character in keeping with Hollywood's trend of making all characters as complex and layered
as possible. While it's difficult to claim that scientists today are simply nerds
or villains, those archetypes have not entirely gone away either, and are the
first two represented.

The Mad, Evil, or Power-hungry Scientist This is the scientist who
strives to uncover "forbidden" knowledge that is too dangerous for humanity
to possess, or is amoral, unethical, or in service to his goal of global domina-
tion. Dr. Victor Frankenstein would be the prototypical example. Although
this archetype has occurred far more often in Hollywood past, it still crops up
from time to time. In fact, as the series *Fringe* followed timelines in two parallel universes, Walter Bishop represented two variations of the mad scientist
archetype in a single series.

Another character who, like Bishop, has a Mengele-like past, is the town
doctor in the series *Defiance* (2013–), Meh Yewll. With the war between the
Votanis Collective and Humanity over, she has taken a turn to the good side[74].
Dr. Totenkopf[75] of *Sky Captain and the World of Tomorrow* (2004) would also
be a classic mad scientist, but even though the character appeared in a fairly
recent film, one intent of the filmmakers was to capture the look and feel of

---

[73] Some would claim that there are four with two of lesser, or emerging, significance. We're going with six.

[74] Or has she?

[75] Literally translated, "totenkopf" means "dead head" (though we doubt he was a fan of Jerry Garcia's), but that's actually the German term for a skull and crossbones. Ironically, or perhaps aptly (certainly creepily), Totenkopf was portrayed by the late Sir Laurence Olivier, who had passed away fifteen years prior. The filmmakers used archival footage of Olivier to create Totenkopf.

1930's and 1940's comic books—a time and genre where mad scientists were common denizens.

A show where mad scientists still appear with some regularity is *Doctor Who*, with recent examples being Dr. Lazarus[76], Davros, creator of the genocidal Daleks (very much modeled on Mengele, given that the Daleks were created as analogs to the Nazis, World War II still being fresh in the minds of many Britons when *Doctor Who* debuted in 1963) along with the Doctor's long-time arch-enemy, The Master[77,78].

This archetype was so overused in the past that time has relegated it largely to the realm of comedic films and parody. Recent examples exist more for comedic value than antagonistic: Dr. Evil in the Austin Powers series of movies, and Dr. Horrible in *Dr. Horrible's Sing-Along Blog*[79,80].

The Socially Inept, Absent-Minded, or Nerdy Scientist This character is Cameron's "idiosyncratic nerd": he is the absent-minded professor, the nutty professor, the scientist who shrunk the kids, and every character on *The Big Bang Theory*. This character is often supremely dedicated to his or her work, forgoing non-scientist friends, family obligations, and anything that smacks of romantic entanglement. More true-to-life examples would be Gil Grissom in *CSI:Crime Scene Investigation* television series, Ian Malcolm (Jeff Goldblum) from *Jurassic Park*, or Dr. Roger Kadar (Robert Sean Leonard) in *Falling Skies* (2011–2015).

The Scientist as Sidekick Hollywood is depicting scientist characters increasingly frequently in the role of sidekick:the trusted compatriot who does the research—the digging, the fact checking, the number crunching—that allows the hero to focus on the job of being, well, heroic. Prototypical examples of this archetype might be Dana Scully (Gillian Anderson[81]) in the *X-Files* and Mr. Spock (Leonard Nimoy) in the original *Star Trek* television series. This archetype is both common and useful on television today, and is probably the archetype most closely linked to the "exposition dumper" of years past. Lucius Fox (Morgan Freeman)—the scientist and inventor from Wayne In-

---

[76] Dr. Lazarus was played by the insanely talented Mark Gatiss, who has also written several episodes of *Doctor Who*. Seriously, check this guy out on IMDb, he's *en fuego*.
[77] The Master was originally introduced in the 1971 episode "Terror of the Autons," and Davros first appeared in the 1974 episode "Genesis of the Daleks." So, in some sense, these examples still belong to the cinematic past—the characters recur because they are long-time fan favorites.
[78] Or, as of the 2014 episodes "Dark Water" and "Death in Heaven", the Mistress (aka "Missy").
[79] He has a Ph.D. in horribleness, after all.
[80] We must withhold judgment on Bad Horse—the question remains open whether his evil death whinny is a superpower or achieved through technological enhancement.
[81] Who also drives a lot of the exposition/backstory as Dr. Bedelia Du Maurier in *Hannibal*.

ho supplies Batman with his array of cool tech toys[82] in the Batman —qualifies. This character is essential, and is a fixture, in most crime shows—the CSI crime scientists in the popular television series *CSI:Crime Scene Investigation* (2000–), Abby Sciuto (Pauley Perrette) in *NCIS* (2003–), or Dr. Beverly Katz (Hetienne Park)[83] and the comedy duo of Price and Zellar in *Hannibal* (2013–).

The Scientist as Powerless Pawn In comparison to the mad scientist, this scientist is often more of a passive villain in the service of a mad scientist, a large corporation, a wealthy industrialist, or the military. Although many scientists are driven to be at the forefront of human knowledge, they may not be able to afford to assemble high-tech devices from their research grants. So although scientists may have the knowledge to create cutting-edge technology, it is the large corporation, the military, or the wealthy industrialist who has the financial capital to make this happen. The graduate students in the film *Real Genius* (1985)—who labor under an unscrupulous professor to construct a high-powered satellite-based laser whose application is targeted assassinations—would qualify[84]. Dr. Vargas, Dr. Jennings, and the other scientists conscripted to work for Dr. Totenkopf in *Sky Captain and the World of Tomorrow* are excellent examples, as are the scientists working for John Hammond's InGen in *Jurassic Park* (1993). One could argue that Lucius Fox belongs in this category, but the implication is typically that the corporation, investor, or military for whom the scientist is in service is evil or corrupt. This is not the case with Wayne Industries.

The scientists in the 2000 film *Hollow Man* skirt this category in interesting ways, particularly Linda McKay (Elizabeth Shue), Sebastian Caine (Kevin Bacon), and Matt Kensington (Josh Brolin). While the scientists in this film are performing invisibility experiments for the military, due to a well-placed fib (that they are behind schedule when, in fact, they are ahead of schedule) by team leader Caine, the team of scientists are working less on the military's schedule, and more on Caine's. In that instance, most of the scientists are still pawns, but pawns of Sebastian Caine not the military. After Caine tests the invisibility serum on himself, one side effect is psychosis, thus putting him firmly in the category of mad scientist by the film's half-way point.

---

[82] Take *that*, James Bond, you're not the only one with a Q.

[83] SPOILER ALERT: In the episode "Mukozuke," Beverly becomes... Katzup.

[84] It could be argued that these characters fall under the nerd category, but rather than being portrayed as toiling in single-minded obscurity, these characters were depicted has having active social lives—even if it was with other nerds.

The Scientist as Hero This modern evolution of the scientist characters represents two great tastes that taste great together. It's Kirk and Spock after the transporter accident. It is the scientist as protagonist—in the lead role or even as action hero. It's Jack Hall (Denis Quad) in *The Day After Tomorrow* (2004). It's Alan Grant (Sam Neil) and Ellie Sattler (Laura Dern) in *Jurassic Park*. It's Ellie Arroway in *Contact*. It's Cliff Buxton (Sam Neil again) in *The Dish* (2000)[85]. Historical examples would be scientists in *The Andromeda Strain* (1971). Dr. Heywood Floyd (Roy Scheider) in *2010: The Year We Make Contact* (1984), the scientists in *Time Travelers* (1964), and ... and ... did we mention *The Andromeda Strain*? This example has been historically somewhat rare, but is now increasingly common. Perhaps the Hollywood Curriculum Cycle is proving that scientists do make believable and compelling protagonists. Spock, in the rebooted *Star Trek*, has, by narrative necessity, eschewed the role of "explainer," and assumed more of an action hero status on par with Captain Kirk.

The Scientist as Conflicted Prantagonist[86] Just because your narcissism and libido result in the nuclear incineration of 50 billion people, that doesn't make you a bad person, does it? We speak, of course, of Gaius Baltar (James Callis) from the reimagined *Battlestar Galactica*, bringing us to the final archetype: the scientist as deeply flawed, or morally ambiguous, prantagonist. In the original 1978 *Battlestar Galactica*, Baltar[87] was a greedy power-hungry politician who sold out humanity to their centuries-old enemy—the robotic Cylons[88]—for the promise of power[89]. When first we meet Dr. Baltar in the reimagined series, it is clear very early on that he is narcissistic, self-serving, and hedonistic, yet over time he demonstrates that his poor judgment, dubious decisions, and astoundingly questionable dating preferences haunt him (Fig. 3.5). Similar to Gaius Baltar, *The Core*'s Dr. Conrad Zimsky (Stanley Tucci) is brilliant, narcissistic, and not beyond scientific "duplication of results with insufficient reference."[90] Zimsky is also charming, confident and, when the *Virgil* mission to save Earth reaches a critical juncture, he gives

---

[85] *The Dish* is actually Australian, but it is included here because of its sheer awesomeness. It's funny, charming, and based upon historical events surrounding the Apollo 11 moon landing. If you've never seen it, take our word for it and *see this film*. You'll thank us for it later. No, really, see it!

[86] Prantagonist is a portmanteau, a mash-up of protagonist and antagonist, just as a combination friend and enemy has been come to be called a "Frenemy" (think Hannibal Lecter and Will Graham).

[87] Baltar had only one name in the original series.

[88] In the original *Battlestar Galactica*, the robotic adversaries had not been created by humans, but rather by an alien reptilian race called Cylons. In a scene deleted from the original broadcast, Apollo explains to his future stepson Boxey that the Cylons built a race of robots who had risen up and exterminated their reptilian creators, and the name "Cylon" was something of a misnomer and a holdover.

[89] Does that scenario ever end well?

[90] Which is a tactful way academics accuse somebody of plagiarizing the work of another.

his life in a noble act of self-sacrifice so that the mission will succeed.

Zack Stentz shares that, on *Fringe*, he and Miller saw the potential of evolving Walter Bishop into a character beyond the mad scientist archetype: "What we were really pushing towards is embracing the implications of the Walter Bishop character before he had … chunks of his brain removed. That he had been a terrifying Mengle-like figure of scientific brilliance, but utter amorality, before he became the kind of loveably addled figure of the present day."

A standout example of how far scientists have come in their Hollywood representation is the character of Dr. Becca Kelly (Lili Bordán) in the recent *Battlestar Galactica* prequel series *Battlestar Galactica: Blood and Chrome* (2012). A computer scientist for Graystone Industries, Dr. Kelly is the kind

Fig. 3.5 Conflicted? Gaius Baltar? He's a prototypical scientist prantagonist. James Callis is Gaius Baltar in the reimagined *Battlestar Galactica*. Copyright © British Sky Broadcasting, David Eick Productions, NBC Universal Television. Image credit: MovieStillsDB.com

of competent, strong, confident, female character for which *Galactica* developed a reputation. As the series unfolds, and as the plot wends through the kinds of twists and turns[91] viewers have come to expect from *Galactica*, Kelly appears, at various times, to serve in the role of sidekick, corporate pawn, prantagonist, action hero, and mad[92] scientist. In addition to being a female scientist in a leading role, the character of Dr. Becca Kelly is as complex and layered as her male counterparts—a long way from the one-dimensional exposition dumper of years past.

3.6 Is Asperger's the New Lab Coat?

Screenwriters have many mantras by which they live, one of the most important being: "Show, don't tell." If something is important enough to get into dialogue, it is important enough to get on the screen. Images often convey in-

---

[91] That make this prequel worthy of the moniker *Battlestar Galactica*, in our humble opinions.
[92] Or, at least, extremely peeved.

formation, emotion, and drama more powerfully, and cert
than do spoken words. Words and imagery together can
to set a mood, or create an emotional viewing experience
ful than either alone. Words also come at a premium in a s
in a modern teleplay when, for an hour drama, the writer
to tell a story that has a beginning, a middle, and an end[93]. In much the
same way that screenwriters "hide" exposition, if they can convey information through visual clues, as opposed to spending precious screen time on
dialogue, then visual clues are the route the screenwriter will take.

So to identify a character as a scientist, imbued with knowledge and credibility for later when they must deliver exposition, early on another character
may refer to the scientist character as "Doctor" or "Professor." A shorthand
just as effective, though, is to introduce the character wearing a lab coat, in
a laboratory, solving equations at a blackboard, in an office overflowing with
books, rolled up maps, or other symbols of knowledge, or some combination
of all of these. Albert Einstein's shock of unkempt hair was "borrowed" to give
instant scholarly credibility to Professor Jacob Barnhardt (Sam Jaffe[95]) in *The
Day the Earth Stood Still* (1951), and Doc Brown (Christopher Lloyd) in the
Back to the Future trilogy (1985–1990). Jedi Master Yoda's eyes were patterned
after Einstein's to reflect kindly wisdom[96].

Rather than resorting to the lab coat trope, a new method has become
popular in identifying a character as a gifted, but quirky, savant: establish
early that the character has traits associated with Asperger Syndrome—an autism spectrum disorder. "Part of what seems to be happening with Asperger's
now is that it's almost becoming shorthand, kind of the way OCD has," Erika
Drezner, a coordinator for the Asperger's Association of New England, told
Mashable. She adds, "It means they're nerdy and smart and not very socially
gifted."[97] Zack Stentz, who, with two children of his own on the autism spectrum, agrees, "The socially maladroit/aspy scientist has become a bit of a cliché. It went from being not very common to, kind of post *Big Bang Theory*,
almost the default place that you go to."

Approximately one in 500 Americans has Asperger Syndrome, which, by
coincidence, is approximately the same percentage of people who are Ph.D.-

[93] Along with five act breaks.

[94] KRG tried desperately to get two words of sciency/expositiony text into the second season *Battlestar Galactica* episode, "Valley of Darkness," and there simply wasn't room. That's how tight a modern-day television script can be. In case you're interested, the words were "integrate backwards".

[95] In the 2008 remake, Professor Barnhardt is played by Monty Python alumnus John Cleese.

[96] Hauptfuhrer F, Peterson K (1980) Yoda Mania: America Falls in Love with the 26–Inch, Green, Pointy-Eared Sage and his Master Puppeteer, Frank Oz. *People*. Retrieved 28 July 2014.

[97] When Hollywood Turns Asperger's Into a Joke: http://mashable.com/2014/05/08/aspergers-tv-movies/

ing scientists. Although there is not a one-to-one correlation between the two populations, the overlap is not insignificant. Asperger Syndrome, or simply Asperger's, is an autism spectrum disorder that gets its name from the German pediatrician who first noted the condition in 1944.

The modern concept of Asperger's, as a disorder with enough specific traits to warrant its own diagnosis, came into being in the 1980s and early 1990s. Common behavioral traits include average to above average intelligence. Those with Asperger's are frequently socially impaired because they are unable to empathize with others, and/or fail to recognize emotions or facial expressions in others. They are also typically obsessive, rigid, and averse to change. While the term "Asperger's Syndrome" has been replaced by updated terms in the latest revision of the Diagnostic and Statistical Manual of Mental Disorders, the DSM-5, it is a term that rapidly conveys meaning in colloquial setting. So like the terms "psychopath", "insanity", "right-brained", and "left-brained"—terms which have long been abandoned by mental heath care professionals—the term "Asperger's Syndrome" will likely be with us for some time, particularly in TV and film.

Recent EEG research[98] suggests that, although autism and Asperger's share similar connectivity in certain parts of the brain, as well as many of the same behavioral manifestations, there are also physiological variations in brain connectivity that distinguish brains with autism compared to brains with Asperger's. These physiological differences may help to explain developmental differences, and why children with Asperger's do not have the cognitive or language delays that are typically associated with autism.

Frequently labeled "nerds" or "geeks" by others, Aspies, as they even refer to themselves[99], are known for having one, or sometimes a few, intensely focused interests—which range from the mundane to the unusual. Even if the "special interest" is not out of the ordinary, the intensity of the focus on the interest is. Some Aspies like *Doctor Who*, some like baseball cards, some like pet rats.

Some like science.

In fact, it is fairly often that the special interest is science- or computer-related. Diane Kennedy, author of *The ADHD-Autism Connection: A Step Toward More Accurate Diagnoses and Effective Treatment* writes, "They are our visionaries, scientists, diplomats, inventors, chefs, artists, writers and musicians. They are the original thinkers and a driving force in our culture."[100]

---

[98] Asperger's and Autism: Brain Differences Found: http://www.livescience.com/38630-autism-asperger-eeg-connectivity.html
[99] People who are not on the autism spectrum are referred to as *neurotypicals*, or NTs.
[100] Kennedy D (2002) *The ADHD-Autism Connection: A Step Toward More Accurate Diagnoses and Effective Treatment*. Waterbrook Press, Colorado Springs, CO

Hans Asperger believed that "…for success in science or art, a
is essential."

> Autism is cool. When I like something, I like it a whole lot more than my friends like their things.
>
> Anonymous Aspy

Formally identified or not, scientist (or related) characters with Aspy traits are commonplace in Hollywood today—examples are on television with Temperence "Bones" Brennan, Gregory House, Adrian Monk, and Gary Bell of *Alphas* (2011–2013)—all of whom display behavioral traits of Asperger's. Nowhere is a character with Aspy traits taken to more extreme than in the 2004 film *Raising Genius*. Hal Nestor (Justin Long[101]) is a mathematical genius who lives locked in a bathroom, and spends the day deriving equations[102] on the walls. Obsessive behaviors (with no particular genetic link to one another in the real world) run in the family: Hal's father has a hoarding disorder, and his mother, eager to prove to the world she's a good parent, is obsessed with Hal leaving his bathroom to give his high school valedictorian commencement address. Reviews and synopses of the film claim that Hal is highly obsessive-compulsive, but his social impairment, complete lack of empathy, narrow range of interests, lack of social reciprocity, prosidy, and language quirks all argue much more strongly in favor that Hal is Aspy. The film is difficult to get through; none of the characters are written to be particularly likeable or empathetic, but the viewer is left with a glimpse of what life is like living with somebody with a more extreme manifestation of Asperger's than is typically shown in Hollywood.

Adam (2009) explicitly explored the day-to-day life and struggles of a man with Asperger's[103]. Hugh Dancy plays Adam Raki, a man who works as an electronics engineer for a toy company in New York City. After the death of his father and the loss of his job, Adam develops a relationship with Beth Buckwald (Rose Byrne), his neighbor upstairs. At various times, Adam and Beth bond over Adam's special interests—his love of astronomy and space exploration, and his fascination with a family of raccoons living in nearby Central Park.

---

[101] Long was, famously, the Mac in the old PC/Mac commercials (opposite John Hodgman who was the PC). In the film, his parents give him a computer as a graduation present and, naturally, it's an Apple.
[102] "Formulas!" Nestor insists … but "equations" is the better term.
[103] Max Mayer, who both wrote and directed *Adam*, won the 2009 Sundance Film Festival's Alfred P. Sloan Prize.

Fig. 3.6 Adam bonds with his neighbor Beth over the latest images from the Cassini spacecraft orbiting Saturn in *Adam* (2009). Copyright © Fox Searchlight Pictures. Image courtesy MovieStillsDB.com

Ultimately, Adam receives a job offer to design astronomical electronics in Flintridge, CA which, given his abilities and interests, seems to be a perfect fit for him. Since Adam does not mention the potential employer initially, the viewer may reasonably conclude that the offer is from NASA's Jet Propulsion Laboratory[104] (autistic scientist and autism advocate Dr. Temple Grandin[105] once referred to NASA as the "largest sheltered workshop in the world" for people with autism and Asperger's). A year later, Adam is happy in his new job, which actually turned out to be at Mount Wilson Observatory, with the famous Hooker Telescope[106] even making a cameo appearance (Fig. 3.6).

Another prototype for the Aspy scientist is *The Big Bang Theory's* Dr. Sheldon Cooper. On whether or not his character has Asperger's, in an interview with *Variety* actor Jim Parsons shared, "The writers say no, he doesn't … [But] I can say that he couldn't display more facets of it."[107,108] Zack Stentz adds, "Twenty years ago pretty much all you had was *Rain Man* (1988), that was all people knew, and so the idea that characters on the spectrum can be function-

---

[104] Although reporters will say that JPL is based in Pasadena during broadcasts, the lab is physically located in La Cañada Flintridge, CA.

[105] In 2010, HBO produced a biopic entitled *Temple Grandin*. Grandin was played by Claire Danes, who won a Golden Globe, an Emmy, and a Screen Actors Guild Award for her performance.

[106] The Hooker Telescope is the telescope on which Edwin Hubble did his ground-breaking work on galaxies and the expansion of the Universe. More on him later.

[107] 'Big Bang Theory': Jim Parsons—'Everybody has a little Sheldon in them': http://variety.com/2008/tv/news/big-bang-theory-3-27173/

[108] In *The Big Bang Theory* episode "The Closure Alternative", Sheldon Cooper is so distraught after *Alphas* is cancelled following a cliffhanger ending, that he tracks down the writer of the finale, Bruce Miller, to learn how the series-ending cliffhanger would have been resolved. Miller is a real person with whom KRG worked on *Eureka*.

al and have gifts and even be cool in their own way … *The Big Bang* ⟨ is, in my mind, a step forward."

> For all we know, the first tools on Earth might have been developed by a loner sitting at the back of the cave, chipping at thousands of rocks to find the one that made the sharpest spear, while the neurotypicals chattered away in the firelight.
>
> Dr. Temple Grandin, Animal Scientist/Autism Activist

Stentz who, along with Miller[109], wrote for *Terminator: The Sarah Conner Chronicles* (2008–2009) says, "You know what's really funny is, we actually modeled a lot of Cameron—Summer Glau's character—on Asperger's traits. Specifically, some of them were from observing my daughter, and to the point where a lot of people in the community kind of glom onto the Cameron character the way that, before that, they glommed onto Spock and Data in the *Star Trek* shows as kindred spirits."

While it is impossible for a concrete diagnosis of Asperger Syndrome without face-to-face interaction—psychologists even have a difficult time distinguishing between Asperger's, attention deficit disorder, and intellectual giftedness—historical records recounting the behavior of many famous scientists have revealed many Aspy-like traits, and fueled speculation that they may have had Asperger's: Benjamin Franklin, Marie Curie, Paul Dirac, Alan Turing, and Charles Richter, to name a few.

There have never been many scientist role models. Isaac Newton, Nikola Tesla, and Albert Einstein were all world-famous and had considerable "cults of personality" surrounding them. Based upon historical records of their personalities there is speculation each—the scientist role models of days past—had various Autism Spectrum Disorders like Asperger's. So Hollywood is really doing nothing new, it is simply the case that Asperger Syndrome was formally recognized only fairly recently. Asperger's is not the new lab coat, it is the *old* lab coat, and yesterday's "idiosyncratic nerds" are today's Aspies.

> Mild autism can give you a genius like Einstein. If you have severe autism, you could remain nonverbal. You don't want people to be on the severe end of the spectrum. But if you got rid of all the autism genetics, you wouldn't have science or art.
>
> Dr. Temple Grandin, Animal Scientist/Autism Activist

---

[109] Miller and Stentz also penned a critically acclaimed young adult novel about a high schooler with Asperger Syndrome entitled *Colin Fischer*, as well as the sequel *Colin Fischer: Moore, R.T.*

3.7 Science as a Superpower

Although scientists are, by and large, being depicted more positively by Hollywood today, this does not necessarily mean that they are always depicted accurately. Biographical films (often called biopics) that purport to chronicle the lives of real scientists are promoted as telling a "true story," but even the best biopics rely upon significant artistic license and notable omissions. Furthermore, while scientists lament at length about some stereotypes, there are other clichés that go largely unchallenged, because they cast scientists in a flattering light. For example, on *Eureka*, most of the ensemble cast—as well as many of the recurring characters—were portrayed as unfathomably competent. Detractors of the *CSI* franchise have pointed out that the show's characters not only process crime scenes, but also carry weapons, pursue and interrogate suspects, conduct raids, and solve cases—all of which is the job of detectives or uniformed law enforcement, not forensic workers[110]. Yet this cliché of universal expertise can be damaging in its own way to scientists, because it creates impossible-to-fulfill expectations.

For shows like *Eureka, Fringe*, the various incarnations of *CSI, Star Trek* and *Stargate*, and many others, universal expertise arises for both dramatic and budgetary necessity: no television series, no matter how well-funded, has the capital to afford dozens of different, though realistically competent, characters, and the audience does not have the emotional capital to follow and maintain interest in that many characters. Some shows do frequently bring on additional expert scientist characters for single episodes, multiple episode story arcs, and even half- or season-long arcs, but, as Stentz points out, "There's the perennial problem that you don't want to give all the good stuff to the guest artists."

Bragi Schut, who has written for both television and film[111], shares another good reason why screenwriters might choose to make their scientist characters "the best of the best."

> The thing that I'm writing right now has scientists in it, and they're sort of the "best in their fields." They've been brought together to fight this impending sort of doom that's coming, and they've got to find a solution. I think that some of that stuff is just by necessity, because in this story … it had to feel

---

[110] This is not always a legitimate complaint, in some police departments, the CSIs do, in fact, do all these things (generally because smaller departments can't afford to hire dedicated forensic specialists). So although the criticism may be valid in Las Vegas, home of the original *CSI: Crime Scene Investigation*, it is not valid in *CSI: Miami*. For a more detailed examination, see Cass et al. in *Hollywood Chemistry: When Science Met Entertainment* pp 145–151.

[111] Schut also created the television series *Threshold* (2005).

enormous, and that we were throwing billions of dollars at it, and the best minds were brought together to fight this thing. So we try to populate the script, and bring characters in, to face these threats that feel credible, and real, and that are really good at what they do. You want to see people come forward with the best plans, and they still barely pull it off by the skin of their teeth.

When universal expertise is taken to its ultimate extreme, scientists are portrayed as infinitely capable in all subject areas. In the original *Star Trek*, how often was Spock out of his scientific depth on any topic? How often does a spaceship-bound scientist character boast, "I know every inch of this ship like the back of my hand, I *designed* her," when a more realistic situation is that a scientist or engineer might be in charge of multiple teams of scientists, engineers, and technicians, each of which designs a single subsystem, instrument, or even a lone integrated circuit chip? (For a real life example, look at the invention of the atomic bomb: a team effort by some of America's greatest minds, under the leadership of Robert Oppenheimer.) Stentz says, "I think a lot of it, especially in more episodic television, and in a film that has to be resolved in two hours, is there is always the temptation to make your scientist your MacGyver—essentially one or two characters that you go to solve everything by the end of the hour."

Ironically, then, universal expertise can be as limiting from a dramatic standpoint as it is enabling, as Ashley Miller explains:

> Contending with failure, or perceived failure, or setback is the heart of drama, and there could be something very powerful about allowing scientists to be incredibly smart, incredibly competent, but still, like everybody else, if they are batting .300 they're doing really well. The freedom to be wrong is an amazing gift to give any character, and we deny it to scientist characters much of the time because we are leaning on them to move the plot forward.

Clearly there are many good reasons why, like nerd and mad scientist, Hollywood should humanize the übercompetent scientist as well.

3.8 Scientist Representations II: Bigger, Badder, and Better Than Ever

Although Hollywood got off to a rocky start where scientists were concerned, the depictions of scientist characters are now as intricate and layered as that for teachers, trial attorneys, or Targaryens. Producers and writers are also aware that the same demographics that make real-life scientist role models extremely scarce, also makes them a very minor audience segment in the cut-throat competition for box office return and Neilson ratings[112]. Apart from a dedication to making all characters, especially in ensemble casts, interesting, complex, and compelling today, Hollywood does not view scientists as a loud, vocal minority to which they must cater.

The showrunner for science fiction series *Defiance* and *Caprica* (2010), Kevin Murphy believes: "I'm sure that there are lawyers whose heads just spin around watching a great law show like *The Practice*. I'm sure doctors watching a fantastic medical show like *House* are tearing their hair out because, 'Urgh, that's *not right!*' And I'm sure scientists watching any science fiction show have a real sphincter-clenching adverse reaction to what they're seeing, because they know too much about the world. But ... scientists are not our core audience. Our core audience is everyone else."

Irrespective of why scientist portrayals are improving, they are improving. Today for every *WoW*-playing, Comic-Con-attending, stereotypically nerdy Douglas Fargo, there is a handsome, buff, motorcycle-driving, roguish computer genius like Zane Donovan[113]. Media experts Matthew C. Nisbet and Anthony Dudo, who have studied how science and scientists are portrayed in the media, share:

> Since as early as the 1970s, many members of the scientific community have criticized entertainment, television, and film portrayals for promoting negative stereotypes about scientists, for featuring improbable or inaccurate scenarios and depictions, and as contributing to a perceived culture of "anti-science..." Contrary to the fears of many scientists ... research in fact suggests that over the past decade there has been a trend towards ever more positive "hero" portrayals of scientists.

Yet Murphy believes that populating worlds like Caprica and towns like Defiance with believable scientist characters is an important component for making science fiction shows more immersive for viewers, "Better science makes better story because it makes it easier for you to believe that these people really know what they're talking about, and they know more than you do about the given subject, and you trust the storytelling, and you find yourself invested and, again, it all comes down to the word *verisimilitude*: I believe in the world."

So, we would make a plea to scientists and engineers: before getting your hackles up the instant one of your sci-fi counterparts appears on screen, take a

---

[112] Besides, scientists don't tune in to much television anyway, they're too busy trying to cure cancer, end global warming, probe the origin of the Universe, and feed the world, to watch much television.
[113] Both Fargo and Donovan appeared on the SyFy Channel series *Eureka*.

moment to see if you're really being portrayed all that badly—think of all the lawyers, doctors, spies, and cops you've enjoyed watching and ask yourself if you're really being treated any worse? Better yet, imagine what the ten-year-old version of yourself would think of the character, before you learned the hard way about the "99 percent perspiration" part of doing real research. Be honest—we bet you'd think that character was pretty cool!

These days, Mr. Cameron might, too.

> The happiest people I've found are in science. They love what they are doing, they have a good family life, they're satisfied.
> Eli Broad, Entrepreneur/Philanthropist

4
Matter Matters

Energy is liberated matter; matter is energy waiting to happen.
Bill Bryson, *A Short History of Nearly Everything*

Sir, may I recommend the barium hydrochlorate salad nicoise, followed by the helium-3 isotopes de la meson, and then perhaps a small radioactive fruit salad for pudding?
Arnold J. Rimmer, *Red Dwarf,* "The Last Day"

The ships hung in the sky in much the same way that bricks don't.
Douglas Adams, *The Hitchhiker's Guide to the Galaxy*

Matter, energy, and radiation are the fundamental subjects of physics. They're also the raw material for endless plot points in science fiction TV shows and movies, giving superheroes their powers, providing the fuel to send spacecraft through hyperspace, and creating unbreakable swords and impenetrable armor. Understanding the concepts of mass, energy, how energy affects mass, how different forms of energy are converted from one to the other, and how matter and energy are converted from one to the other, helps a scientifically-armed viewer understand what's possible and what's outlandish in a screenplay.

A key part of the story of matter, energy, and radiation is that they are all, in a sense, the same thing: they are different aspects of the same fundamental material of existence, and quantum mechanics is the set of rules by which they exist. Physicists believe that a fuller understanding of the complex subtleties of these aspects would lead to a deeper understanding of universe—maybe even to an ultimate Theory of Everything that can explain all phenomena in the cosmos, from how the atoms of our bodies behave to the fate of galaxies.

Because they are different manifestations of the same thing, any discussion of matter, energy, and radiation rapidly leads to a "chicken and the egg" scenario: to describe matter, the concept of energy quickly enters the picture. A discussion of radiation requires an understanding of both matter and energy. To understand all three requires a basic understanding of quantum mechanics, so it is less of a chicken and the egg scenario, and more of an Alien queen, egg, face-hugger, chest burster[1] scenario. It's probably easiest to start with matter—for no other reason than is the topic that is the most intuitive and familiar in our everyday lives.

Neophyte physics students are counseled that they will "learn everything three times": the basic version, an advanced version, and the "correct" version. Perhaps there is no better example than the example of the concept of mass, a property that all matter possesses.

4.1 Mass: The Sum of All Nucleons

In its most basic sense, mass can be viewed as a tally of the amount of matter an object contains—a count of the total number of electrons, neutrons, and protons that compose an object. In fact, even though there are as many electrons as protons, electrons represent a negligible contribution. Mass is commonly measured in kilograms: a kilogram was originally defined as the mass of water in a volume of one liter at $4\,°C$ (the temperature at which water is at its densest, which is why colder ice floats). Today one kilogram is defined by reference to a block of 90 % platinum/10 % iridium alloy in a climate-controlled vault of the International Bureau of Weights and Measures[2] in the town of Sèvres on the outskirts of Paris. The official kilograms in the vault (as well as the reference block, there are six copies), are all cast in the form of cylinders with equal height and diameter of 39.17 mm. (Unfortunately, the reference blocks grow slightly in mass over time as they pick up stray atoms from the atmosphere and have to be periodically cleaned, so the search is on to find a way to define the kilogram in terms of fundamental physical constants, which is how other key units, such as the second and the speed of light, are defined.)

Our understanding of the basic building blocks of matter dates back to the ancient Greek philosopher Democritus, who noted that when a stone is split, the resulting pieces have the same properties as the original rock. If again divided and subdivided, the same properties hold for the smaller pieces. He reasoned that there was a limit to this progression and eventually you would be left with pieces so small that they could not be further subdivided. Democritus named these fundamental pieces "atomos," meaning "indivisible," and believed that atomos were unique—in both shape and properties—to the material that they comprised. For example, he believed that the atomos of stone were solid, interlocking, and unique to stone while the atomos of water soft and slippery and unique to water.

Alchemists, practitioners of the proto-science that led to chemistry and much of modern-day medicine, concurred with the ancient Greeks that a

---

[1] That's mostly a good analogy. Mostly.
[2] Called BIPM by the French—the *Bureau International des Poids et Mesures*.

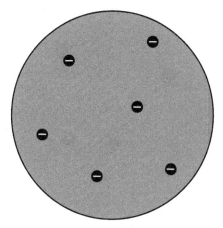

Fig. 4.1 J.J. Thompson's "plum pudding" model of matter

quantity of any substance could, in principle, be subdivided repeatedly until reaching a limit: a tiny indivisible unit called an atom. Alchemists further discovered that the matter that filled the world around them was made of many different indivisible substances, each with different chemical properties, called elements, that went well beyond the classic elements of Earth, Air, Fire, Water, and Aether acknowledged by the Greeks (and the Mondoshawan aliens in the 1997 film *The Fifth Element*).

Today scientists define an atom as the smallest unit of an element that possesses its distinct chemical properties. So hydrogen, helium, carbon, oxygen, iron, gold, and uranium are examples of elements; Wolverine's adamantium, Captain America's vibranium[3], and Pandora's unobtanium are not (in addition to all three being fictional, adamantium is an alloy of several metals, while unonbtanium is a mineral ore). In the late 19th century, and early 20th, scientists discovered that individual atoms had constituent components and, contrary to the beliefs of the ancient Greeks and the alchemists, were in fact divisible. In 1897, experimental physicist J.J. Thompson discovered that a negatively charged[4] particle, which he named the electron—the first-known subatomic particle—resided in every atom. He imagined the atom to be like a plum-pudding: negatively charged electrons were embedded throughout a positively charged material, as in Fig. 4.1.

In 1900, physicists Ernest Rutherford, known as the "Father of Nuclear Physics," and Paul Villard, working separately, identified three different types

---

[3] In the early Captain America comics, Cap's shield is composed of vibranium—in more recent years, it's a vibranium-adamantium alloy. In the films, it is, apparently, a product of Stark Industries.
[4] Charge is a fundamental property of subatomic particles that cause them to experience electrostatic repulsion when in the presence of particles with like charge, or attraction when in the presence of opposite charges.

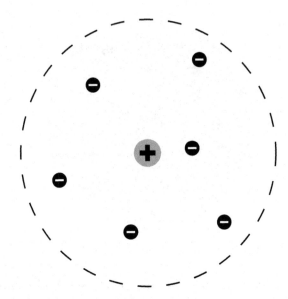

Fig. 4.2   Rutherford's revised model of the atom

of radiation (see Chapter Six, "Radiation: An All-Time Glow"), and named them "alpha rays," "beta rays," and "gamma rays"[5]. Seven years later, Rutherford discovered that alpha rays were actually positively-charged particles, and were identical to the nuclei of helium atoms. Because of this discovery, Rutherford proposed a new model of atoms with negatively-charged electrons surrounding a dense positively-charged particle (which would later be called a nucleus) as in Fig. 4.2.

The Rutherford model was refined by physicist Neils Bohr[6] who proposed that the atom was like a planetary system—the negatively-charged electrons were in orbits about the positively-charged nucleus. This model, the Bohr Model (more correctly, the Bohr-Rutherford Model), is one of the most famous scientific models in existence, its fame on par with the most famous equation $E = mc^2$. It was used at the symbol for the U.S. atomic energy commission, and it is still seen today when we wipe from one scene to another in *The Big Bang Theory* (Fig. 4.3).

In 1917, Rutherford bombarded nitrogen with alpha particles, and discovered that a small positively-charged particle was released. This particle, he learned, was a hydrogen nucleus. Rutherford had learned that the atomic nucleus could also be subdivided, and he is credited with being the first person

---

[5] The very same gamma rays responsible for the genesis of the Hulk. We're getting there.
[6] Rumor has it that Bohr preferred experimental physics to teaching, because students found him Bohring.

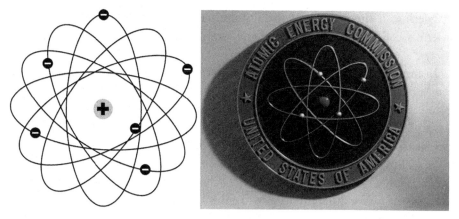

Fig. 4.3 The Bohr model of carbon, and its use on the logo of the Atomic Energy Commission showing just how widespread the model has become. Photo credit: U.S. Department of Energy

to observe how natural radiation could "split the atom"[7], and he discovered a new subatomic particle, which he named the proton.

Finally, in 1932, particle physicist James Chadwick discovered that another type of particle, with roughly the same mass as the proton, also resided in atomic nuclei. Since that particle was electrically neutral, it was called the neutron. Since protons and neutrons both reside in the atomic nucleus, collectively they are called *nucleons*. This was one of many modifications to the original Bohr Model of the atom that would follow. In fact, scientists have known since the mid-20th century that the Bohr Model does not depict the actual workings of an atom very well, but as a tool for teaching basic atomic structure, it survives to this day[8].

Like a planetary system, where the overwhelming majority of the mass of the system is within the central star, the nucleus contains the bulk of the mass of an atom. Planets are bound to their star by the attractive force of gravity; electrically negative electrons are bound to their positively-charged nuclei by electrostatic attraction.

The planetary system as analogy for the Bohr Model of the atom has yet another similarity: the distance between objects is vast. Just as the distance from, say, Earth to the Sun is much larger than the size of either Earth or the Sun, the distance from the nucleus of an atom to its closest electron orbit is

---

[7] The first people to split an atom using technological means were John Cockcroft and E.T.S. Walton in 1932, for which they received the 1951 Nobel Prize in physics. This breakthrough led to both nuclear energy, and nuclear weapons.

[8] A brilliant example of this, one every educator should see (it's on YouTube), is in an episode of *WKRP in Cincinnati* entitled "Venus and the Man". The character Gordon Sims (Tim Reid), whose on-air name is Venus Flytrap, paints a picture of the Bohr atom—equating sub-atomic particles to street gangs called "The Pros," "The New Boys," and "The Elected Ones"—that has to be seen to be appreciated, and is unforgettable. Hollywood rarely provides such effective "teachable moments" as this.

staggering. If the nucleus of the simplest atom—hydrogen—was the size of a tennis ball then the nearest electron would be 4.3 kilometers (2 ½ miles) away[9].

In most of the matter with which people come into contact in their everyday experience—solids, liquids, and gases—atoms have an equal number of protons and electrons. The positive charges of the protons balance out the negative charges of the electrons leaving most atoms electrically neutral. As chemistry is all about how the electrons of different atoms interact, an atom's chemical properties is determined by the number of protons in its nucleus, a count known as its *atomic number*. Different elements are defined as such because they have different atomic numbers. Hydrogen, one proton with one electron, comprises over 90 % of the visible mass in the Universe. Helium has two protons, lithium three, carbon six, oxygen eight, and so on.

When an atom has an imbalance between its number of protons and electrons, and has a net charge, it is called an *ion*. An ion can be positively-charged or negatively-charged, but it is more common that electrons are stripped off an atom, leaving a net positive charge.

There are many processes that can strip electrons from their atoms, chief of which is simple collisions between atoms. Whack an electron hard enough and it can be knocked off the atom it is orbiting. When a gas is heated to the point where all the atoms are ionized, where the substance has transformed from a random assortment of atoms, to a random assortment of electrons and positive ions (aka atomic nuclei), we call that state of matter a *plasma*. While it is tempting to think of plasma as "just a hot gas," it is as different from a gas as a gas is from a liquid, and as a liquid is from a solid. Although this is actually the most common state of matter in the visible Universe, because it's what stars are composed of, it is the state of matter with which people are least familiar, because the planets are simply not hot enough to produce appreciable amounts of plasma. The form of plasma with which the average person is most familiar is lightning, though neon lights and some TVs also rely on plasmas to produce light (Fig. 4.4).

From plasma torches to plasma rifles to fusion reactions to the electro-plasma power system aboard the starship *Voyager* (notice how B'Elanna Torres is frequently referring to Voyager's EPS—electro-plasma system—power conduits), plasma is a fixture in science fiction, whether explicit or implied. We will discuss plasma more in later chapters.

---

[9] It is slightly more complex than that. More on electron orbits later in the chapter, as well as in Chapter Five: "Radiation: An All-Time Glow."

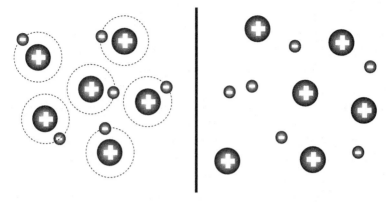

Fig. 4.4 A gas (*left*) consists of atoms with a positively-charged nucleus orbited by a negatively-charged electron. The charges of the nucleus and electrons cancel out, leaving the gas electrically neutral. In a plasma (*right*) electrons are stripped away from atomic nuclei, producing a "sea" of charged particles

> Science is drama. Properly understood, and properly inhabited, there is intense drama inherent in almost every scientific idea. You just have to find it. There's the drama of what's going on, why there's tension between two magnetic poles, or the collision of two particles, or any of the things we talk about. We use the terminology of catalyst and currents and charges and things endlessly in our talk about drama.
>
> Jon Amiel, Director, *Creation*

Add up the number of protons and neutrons of an atom, and that value is the *atomic mass number* of an atom. Some elements can have atoms with different atomic mass numbers but the same atomic number, due to different numbers of neutrons in the nucleus. Atoms of the same element that have a different number of neutrons are called *isotopes* of one another. Let's use the simplest element, hydrogen, as an example. The most common form of hydrogen has a lone proton and no neutrons, giving it an atomic mass number of 1. We call that hydrogen-1, or simply hydrogen. Another "flavor" of hydrogen has one neutron, and we call this hydrogen-2 or *deuterium*.

The number of neutrons in an atom's nucleus has no bearing on the chemical properties of that element. Different isotopes behave the same chemically, more or less. Although deuterium has a little extra baggage, mass-wise, it will still chemically bond with other elements the same way hydrogen does. Instead of having the usual chemical formlua H_2O, heavy water is either HDO[10]

---

[10] HDO is sometimes called "semiheavy" water.

or D_2O^{11}. One interesting curiosity is that although normal water floats, ice cubes can be made from heavy water that sink.

In World War II, the Nazis famously performed "heavy water" experiments that gravely worried Allied scientists. Heavy water can be used to control a type of radiation called neutron radiation in a way that's important to sustaining a nuclear chain reaction, and these experiments were done as precursors to building atomic weapons. Norwegian saboteurs undertook a series of efforts at the plant that was supplying heavy water to the Germans, immortalized in the 1965 film *The Heroes of Telemark*, starring Kirk Douglas and Richard Harris. The Allies also launched a difficult and daring commando raid called *Operation Gunnerside* in 1943 on the same plant. Those efforts, along with allied bombing, meant that the plant never produced enough heavy water to contribute to the Nazi effort to build an A-bomb.

There is never an infinite permutation of isotopes for an element. Some collections of protons and neutrons are so unstable that they split into smaller pieces instantaneously. Some combinations, however, are *metastable*, which means that they exist for some finite period of time, and then spontaneously split, or decay, into smaller nuclei known as *daughter products*—typically emitting some form of radiation in the process.

When an isotope has a metastable configuration that eventually decays naturally, due to no external influence, that isotope is *radioactive*, and is known as a *radioisotope*. For example uranium-238 decays into thorium-234. Daughter products of radioactive decay processes are oftentimes radioactive themselves, and thorium-234 decays into protactinium-234, which decays into uranium-234. After a series of radioactive decays over 4.5 billion years, the final product is lead-206, which is stable.

In our hydrogen example, add a neutron to deuterium, and the result is a third isotope of hydrogen called hydrogen-3 or *tritium*. Recall that it was tritium that powered Doctor Octavius' (played by Alfred Molina) fusion reactor in *Spider-Man 2* (2004). Doc Ock also claimed that there was only 25 pounds of the element on Earth, which is a significant underestimate.

Tritium, the first radioisotope by atomic number, is actually a bit of an oddball. Most radioisotopes decay "down" into daughter products with smaller atomic numbers, as happens when uranium-238 decays. Tritium releases beta radiation, and becomes helium-3, which has a *larger* atomic number, with two protons and a neutron. We said it was an oddball.

Helium-3 has the potential to make fusion power much more practical than at present. Current state of the art in fusion research uses harder-to-handle, but plentiful-on-Earth, hydrogen isotopes as fuel. The He-3 fusion process

---

[11] Water could also be made "heavy" with different isotopes of oxygen as well.

Fig. 4.5 Mission specialist Dr. Ryan Stone and astronaut Matt Kowalski work on the Hubble Space Telescope in *Gravity* (2013). (Copyright © Warner Brothers Pictures. Image credit: MovieStillsDB.com)

is "cleaner"—while most hydrogen fusion processes leave the containment vessel extremely radioactive, the He-3 fusion process does not. The isotope is very rare on Earth, but it is plentiful in other places in the Solar System such as the atmospheres of gas giants such as Jupiter. This is why it often occurs in science fiction as a reason for establishing outposts in those locations. He-3 is also found on the surface of our Moon, and extracting He-3 from the lunar regolith was the reason for the one-man base setting of the exceedingly awesome 2009 film *Moon*.

4.1.1 To Move Mass, Use The Force

In the 2013 film *Gravity*, when the crew of the space shuttle Explorer are given instructions from Houston to return home immediately, the Hubble Space Telescope is still affixed to the spacecraft in the bay. Astronaut Kowalski (George Clooney) disconnects the two, but has a difficult time pushing the Hubble Space Telescope away from *Explorer*. Why when "everybody knows" that the telescope would be weightless in space, was that not a simple task? For that matter[12], the shuttle should be weightless as well, so why didn't both Explorer and Hubble zoom off in opposite directions the instant Kowalski pushed? (Fig. 4.5)

The answer can be found in a second definition of mass: an object's mass is its resistance to a change of motion when a force is applied to the mass. This definition of mass is referred to as *inertial mass*. The more massive an object is, the more subatomic particles comprise its bulk (harkening back to the first definition), and the more force—a bigger push or pull—it takes to move it.

---

[12] See what we did there?

An object's *mass,* should not be confused with its *weight,* though the two are directly related. In casual conversation, we do not generally differentiate between mass (how much material an object contains) and weight (the force of attraction between the object and Earth). The weight of any object is a product of both its mass and the gravitational field acting upon it. (see Science Box "John Carter's Flee Circus").

In 1687 Sir Isaac Newton published the tome *Philosophiæ Naturalis Principia Mathematica* that described his famous three laws of motion—how objects move when under the influence of forces—as well as his Law of Universal Gravitation.

Newton's first law states that every object remains at rest, or at a constant velocity in a straight line, unless acted upon by a force. This is sometimes called the law of *inertia.* A force is simply a push or a pull, large or small: the pull of gravity, the thrust from a rocket engine, the press of a piano key, the explosion in an automobile cylinder pushing on a piston.

Newton's second law states that acceleration is proportional to the force applied, and is, perhaps, the second-most famous equation in science (where F is the magnitude of the force, m is the mass of the object, and a is the amount the object is accelerated):

$$F = ma$$

It explains why it is more difficult to push a stalled SUV than a shopping cart. One has a greater mass to move. Alternately, push a shopping cart with the force it took to move the SUV, and it will take off like a shot. When a Viper pilot is slammed into her seat during launch in *Battlestar Galactica* (either the 1978–1979 or 2003–2009 version), when astronauts and cosmonauts are slammed against chairs and bulkhead by the deceleration of *Alexi Leonov* as it aerobrakes in the atmosphere of Jupiter in *2010: The Year We Make Contact* (1984), and each time a bullet is accelerated out the barrel of a gun by the expanding gases from a small gunpowder explosion, these are examples of forces.

Newton's final law of motion states that for every force there is always an opposing force equal in magnitude and acting in the opposite direction. When Ripley is knocked back by the recoil of her pulse rifle in 1986's *Aliens,* or when Captain Kirk and Spock struggle to hold Doctor McCoy from saving the life of Edith Keeler and altering history in the 1967 *Star Trek* episode "The City on the Edge of Forever,"[13] these are examples of how forces always come

---

[13] Or, in a very similar moment, Jack Carter preventing Henry Deacon from saving the life of Kim Anderson in the 2006 *Eureka* episode "Once in a Lifetime." Despite the show's normally lighthearted tone, this one's a serious tear-jerker. Set tissues to battle stations!

in pairs or *couples*. If you are being attracted to Earth by the force of gravitation, Earth is being attracted to you by the same amount.

Imagine standing on a dolly, perhaps the kind used to service automobiles, holding a shotgun. Each time you shoot the gun, the blast goes one direction, propelling you the opposite direction. Now swap the shotgun for a machine gun. Pull and hold the trigger, and a rapid fire of bullets streams one direction. Each bullet is less massive than a shotgun blast, but there are many more of them. The stream of bullets will, again, propel you in the opposite direction, but in a more continuous, less jerky, way. This is exactly what Quaid/Hauser does in the 2012 remake of *Total Recall*. During a moment when he, and everybody around him, is weightless, Quaid/Hauser uses his machine gun for thrust to propel him down the corridors to escape his pursuers.

Now exchange your machine gun for a fire extinguisher and fire it. The projectiles now are not bullets or shotgun blasts, but CO_2 molecules shooting out of the nozzle in one continuous stream of thrust. It would propel you in the opposite direction of the burst. To fight a fire aboard spacecraft *Mars-1* in *Red Planet* (2000), Commander Bowman applies a burst from a fire extinguisher, the resulting blast sends her flying in much the same manner as Dr. Ryan Stone did in *Gravity*. The difference is that, in *Red Planet*, Bowman was not expecting the kick, and was slammed against the bulkhead, whereas Stone planned on it for propulsion. Later in *Red Planet*, after Bowman vents much of the air aboard *Mars-1* to space to suffocate the fire, she learns that the delta-V[14] imparted from the outrush of air has altered the orbit of the spacecraft around Mars.

There was a fantastic display of all three of Newton's Laws in an episode of the classic *Doctor Who*. In the 1982 episode "Four to Doomsday," the Doctor is trapped in space without a space suit[15], floating untethered between a huge space freighter and his craft, the TARDIS. Running out of air, he pulls a cricket ball from his pocket and throws it against the freighter. The ball ricochets back to the Doctor, who catches it. The momentum of the ball gives the Doctor enough of a kick to get him moving towards his TARDIS, and eventually he's able to make his way inside.

The physics of the scene drew a lot of criticism when the episode first aired. Whenever science is placed "front and center" in this manner, expect a lot of "experts" to weigh in. Here's what would have happened: Ignoring any rotation the act of throwing the ball may have caused, when the Doctor throws the cricket ball, the ball would speed towards the freighter, and by Newton's Third Law, the Doctor would experience the same force and move, albeit much more slowly, to-

---

[14] "Delta-V" is how spacecraft navigators say "change in velocity."
[15] No he would not have frozen, nor would he have popped.

wards the TARDIS due to his greater mass compared to the ball. The ball would undergo a largely elastic collision with the metal hull of the freighter. Assuming it didn't shatter upon impact, the cricket ball would rebound at nearly the same speed it impacted; the spacecraft would feel a very tiny push in the opposite direction[16]. The act of catching the ball would give the Doctor a bit more of a kick, a tiny acceleration, increasing his speed, helping him to reach the TARDIS.

As with many aspects of science on the screen, the rate at which this situation progressed, the speed at which the Doctor moved after both throwing and catching the ball was, perhaps, exaggerated (but there's nothing saying he could not have pulled off the same trick again, but with diminishing returns each time as his speed away from the freighter increases), but as far as the dynamics of the scene, it is a win for physics.

Returning to our *Gravity* example with astronaut Kowalski, this is another cinematic triumph for all three of Newton's laws. When Hubble is first detached, it remains unmoving (it does not simply float away on its own as supposedly weightless objects often do in TV/film). Kowalski gives it a big push, and it begins to move away from Explorer very slowly, because although it is weightless, it is not massless. It takes a big effort to move something that big. Although Kowakski is standing atop *Explorer*, it does not move away at the same rate for, although it felt a force due to Newton's Third Law, the shuttle is far more massive than even the Hubble Telescope, and even that was moving away very slowly with Kowalski's effort.

Science Box: Gravitation and John Carter's Flee Circus

> *Yet although Galileo demonstrated the contrary more than three hundred years ago, people still believe that if a flea were as large as a man it could jump a thousand feet into the air.*
>
> J.B.S. Haldane, Evolutionary Biologist, "On Being the Right Size" (1926)

In Disney's 2012 *John Carter*[17], our protagonist is an Earth man transported to Barsoom, the native name for Mars in Edgar Rice Burroughs's early 20th century pulp novels (Fig. 4.10). In the lesser gravity of the red planet, Carter is able to leap to great heights and over long distances—like a human flea[18]—in order to flee his captors and, later in the film, to gain advantage in battle. The gravity on Barsoom is less than Earth's, but enough to turn man into superman?

---

[16] In reality, the disparity of the mass of the two objects is so great, the energy of the impact would have simply been dissipated in the hull of the freighter as a small amount of heat.

[17] Although this film seems to have run afoul of some influential critics, and was a box office disaster, both authors found it quite enjoyable. While you're waiting for your *Star Wars* fix, why not give this one a try?

[18] Superman has nothing on this guy.

Fig. 4A.1 John Carter (*Taylor Kitsch*) may be able to perform fantastic leaps in the lower gravity of Barsoom, but metal chains are still metal chains. (From *John Carter* (2012) Copyright © Walt Disney Pictures. Image courtesy MovieStillsDB.com)

Sir Isaac Newton, who first attempted to postulate the laws that govern gravity, said that any two objects in the Universe that possess mass attract each other via the force of gravity. The greater the mass of an object, the more of a gravitational attraction it has. Gravitational strength also decreases rapidly, as the square of the distance (r) between two objects increases. Scientists say that gravitation follows an inverse square *law*.

$$F = -\frac{GMm}{r^2}$$

The value is negative to indicate that the force acts in the down direction (down is the direction in which gravitation acts), as opposed to the direction of increasing radial distance between the two masses. The parameter G is known as the gravitational constant[19], and is simply the proportionality constant that allows us to relate the mass of two objects, and the distance between them, with the actual forces that both experience from the gravitational interaction (Fig. 4A.1).

Every two objects that have mass attract each other via gravitation. Even the keyboard of your computer, or a book in your hands, exerts a miniscule gravitational attraction. When Earth and Mars are at their closest, when they are on the same side of the Sun, Mars exerts fifty times the gravitational attraction upon you as a book or a computer keyboard. At their farthest, when they are on opposite sides of the Sun, that drops to a little under five.

An object's *mass* is not to be confused with its *weight*, but the distinction between mass and weight doesn't mean much to the average person. If you multiply an object's mass by the pull of gravity it experiences, you calculate its weight. So mass is the amount of "stuff" contained in an object, weight is the amount of force with which that amount mass presses down on the planet.

---

[19] 6.67×10^{-11} m³/kg·s²

Because the non-astronauts among us live our entire lives in Earth's gravitational field, we easily say that an object "weighs one kilogram", when we should correctly say that the object has the *mass* of one kilogram[20,21].

Digital bathroom or kitchen scales typically have switches to select whether the display will indicate pounds (weight) or kilograms (mass)[22]. Implicit in the calculation of that conversion is the mass being weighed is immersed within Earth's gravitational field. That bathroom scale would be useless on Mars, however, unless you knew the acceleration due to Mars' gravity, and had a calculator handy.

As an example, let's say that Jon Carter's mass is 100 kilograms (kg). He would weigh about 980 Newtons (a measure of force), or 220 pounds, in Earth's gravity. When John Carter is transported to Barsoom, his mass remains at 100 kg, but in the lower Martian gravity, he would weigh less[23].

How much less? Enough to make those fantastic leaps?

Let's start calculating John Carter's weight on Earth by using Newton's Law of Universal Gravitation. The force of John Carter on Earth (F_{JCE}) is given by:

$$F_{JCE} = -\frac{GM_{Earth}m_{JC}}{r^2}$$

Where M_{Earth} and m_{JC} are the masses of Earth and John Carter respectively, and r is the distance between their centers, which is just the radius of Earth. Newton's Second Law says that $F = ma$, but this doesn't look anything like that. Or does it? If we simply re-order the equation above, we get:

$$F_{JCE} = (m_{JC})\left(-\frac{GM_{Earth}}{r^2}\right)$$

So in the above equation,

$$a = \left(-\frac{GM_{Earth}}{r^2}\right) = g_{Earth}$$

We have a special name for this acceleration, one that is even well-known colloquially. The acceleration due to gravitation at Earth's surface is called one g.

---

[20] "We" in this instance, mostly means "Americans." This is an unfortunate side-effect of Americans' refusal to go metric. From an early point in life, American students are taught that 1 kg is equal to 2.2 pounds, obfuscating that one is a measure of mass, and the other is a measure of force. What would be proper to say is that 1 kg exerts 2.2 pounds of force in Earths' gravity. This often makes it frustratingly difficult for teachers and professors who teach basic physics in the U.S. to teach the difference between mass and weight.

[21] SAC, who hails from Ireland where the metric system is in use, believes that the only acceptable place for a pint is the pub.

[22] To further complicate matters, there is also a unit pounds-mass. Few ever use this unit, so for our purposes, "pound" will always refer to a unit of force.

[23] In another link to Mars, a confusion in units of force led to the 1999 loss of the spacecraft Mars Climate Orbiter. Lockheed-Martin, builders of the spacecraft, rated the thruster output in pounds. NASA's Jet Propulsion Laboratory, who operated the spacecraft, had been using metric units for 25 years, and assumed the values were Newtons. As the spacecraft approached Mars, it under-corrected for its final trajectory change maneuvers, passed far too deep within the martian atmosphere, and either broke up or burned up: http://mars.jpl.nasa.gov/msp98/news/mco990930.html.

We've all heard the term g-force, particularly in science fiction when dogfighting pilots experience "...five g's." If we plug in the value for the mass and average radius of Earth into this equation, as well as the gravitational constant, we get:

$$g_{Earth} = \left(-\frac{GM_{Earth}}{r^2}\right) = -\frac{\left(6.67\times10^{-11}\,\sfrac{m^3}{kg\cdot s^2}\right)\left(6\times10^{24}\,kg\right)}{\left(6.371\times10^6\,m\right)^2} = 9.81\,\sfrac{m}{s^2}$$

The value 9.81 m/s² is equal to 32.1 feet/s², the acceleration due to gravity that we all learned in junior high[24].

So, John Carter's weight on Earth is a force, and is the product of his mass and g

$$F_{JCE} = W_{JCE} = m_{JC}g_{Earth},$$

or 980 Newtons. If we transport him to Barsoom, there the mass of Barsoom is only 6.4×10^{23} kg, and its average radius, 3390 km, g_{Mars} falls to 3.72 m/s², or 38 % of Earth's. John Carter's weight is 372 Newtons, or just under 84 pounds. Although, on Barsoom, white men could jump, they could not jump quite like human fleas though, on Jupiter, Carter would get squashed like a bug. At the "surface" of Jupiter[25] he would weigh in at 529 pounds.

In one of those differences that means something only to a scientist, there is a difference between gravity and gravitation. The mutual force felt between two objects with mass is gravitation. Your weight on Earth, what shows on your bathroom scale, is not solely reflected by gravitational attraction. Earth is not a perfect sphere, and some points on the globe are farther from the center than others, causing you to weigh slightly less. Also, the centrifugal force due to Earth's spin actually makes you weigh less depending upon your latitude (most at the equator, least at the poles).

Gravity is the resultant force of gravitation, plus all other forces that add to or subtract from that. The difference between the two is so small that it is unnoticeable to people. Still, the title of Alfonso Cuarón's film would have had less dramatic impact were it *Gravitation*, and even if Mars was rotating twice as fast as it is currently, it still would not have given John Carter the boost he needed to make such fantastic leaps.

Not all forces act on macroscopic scales and involve spacecraft, planets, galaxies, or even cricket balls. There are four fundamental forces of nature that shape how the Universe operates. Two of them, gravitation and electromagnetism, act over potentially infinite distances—across the universe—and mediate everyday phenomena of human experience, while the other two forces manifest on subatomic scales only.

We are all familiar with gravitation—it is what is holding you on your seat, couch, or beach chair at this very moment. All of us have experience with

---

[24] In a voiceover in the second season *CSI:Crime Scene Investigation* episode "Overload," Gil Grissom says, "Terminal velocity [of a falling person] is 9.8 meters per second squared." That is actually the acceleration due to gravity. Terminal velocity, the fastest a person can fall owing to air resistance, is around 56 m/s (200 km/h or 120 mph). To reach terminal velocity a person must fall, roughly, 450 meters.

[25] Jupiter, Saturn, Uranus, and Neptune are composed largely of gas. What is the radius of a ball of gas? If you calculate the pressure of Earth's atmosphere at sea level, then descended into a gas planet to the point where the overlying pressure was the same, THAT defines the radius for that gas planet.

Table 4.1 Fundamental forces

Force	Relative strength	Distance (m)
Nuclear strong	10^{38}	10^{-15}
Electromagnetic	10^{36}	Unlimited
Nuclear weak	10^{25}	10^{-18}
Gravitation	1	Unlimited

different manifestations of the electromagnetic force through phenomena such as static electricity, electric motors, automobile starters, and refrigerator magnets. Our experience with the strong and weak interactions is less obvious in our everyday lives, but the strong force prevents us, our planet, and our entire galaxy from evaporating into quark soup. In fact, although it may be hard to believe if you have ever landed wrong after a slipping on ice or tripping over the dog, but gravitation is, by far, the weakest of the fundamental forces. The nuclear strong force is roughly a hundred billion billion billion billion times stronger than gravity. Table 4.1 shows the relative strength of the four fundamental forces, scaled such that gravity equals 1.

In 1964, physicists Murray Gell-Mann and George Zweig proposed that protons and neutrons were composite particles, and that the atom was divisible yet again, and in 1968, experiments at the Stanford Linear Accelerator (SLAC), determined that each were composed of three fundamental particles called quarks[26]. Quarks are bound together to form protons and neutrons by the nuclear strong force, or strong interaction. The strong interaction differs from gravitation and the electromagnetic force in that it does not gradually decay, the interaction occurs over a finite distance[27], then simply … stops.

Neutrons and protons are also bound by the strong force, so the strong force has to be greater in magnitude than electromagnetic force. In a nucleus, positively-charged protons reside in close proximity to other protons. Through the electromagnetic interaction, like charges repel one another. A nucleus would be impossibly unstable, except that the strong force that binds the quarks together into protons and neutrons "leaks" beyond their boundaries, binding neutrons to protons, and protons to protons. This is what is known as the *residual strong force* (Fig. 4.6).

The nuclear weak force, or weak interaction, allows some quarks to change their nature, in other words it mediates the process of protons turning into

---

[26] *Quark* was the name of a 1977 sci-fi comedy that lasted one season, as well as the name of the bartender on Deep Space Nine.

[27] On par with the size of a proton, whose size was recently downgraded from 0.87 to 0.84 millionths of a billionth of a meter, or 0.84 femtometers.

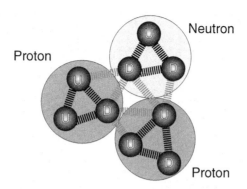

Fig. 4.6   Protons are made of two *up quarks* and one *down quark*, bound together by the strong force. Neutrons have two *downs* and one *up*. The residual strong force binds quarks in different protons and neutrons together to form atomic nuclei.

neutrons, creating unstable isotopes, and enabling certain types of radioactive decay (see Chapter Six, "Radiation: An All-Time Glow").

4.2   Relativistic Mass

In physics the Law of Conservation of Mass says that mass can neither be created nor destroyed. It can, however, be converted into energy, and the most famous[28] equation in all of physics explains how to perform that conversion:

$$E = mc^2.$$

This equation says that when an amount mass (m) is converted into energy, the energy released would be equivalent to the product of the mass times c (the speed of light) squared. The speed of light is a large number; that number squared is a very large number. So a little goes a long way—even a small amount of mass yields a huge amount of energy when converted.

We will use a pair of chocolate bars to put that into perspective. The mass of a typical chocolate bar is 59 grams (2 ounces). Consuming one candy bar would provide you with just over 280,000 calories[29], or about 1.1 million joules of energy[30]. If you were able to use every bit of the food energy your

---

[28] Not to mention surprisingly easy to use.
[29] By an odd quirk of nomenclature, 1 food calorie to a nutritionist is actually 1 kilocalorie, or 1000 calories to a physicist. This goes a long way to explaining why nutritionists are more popular than physicists.
[30] We told you, it's a "chicken and the egg" scenario. See Chapter Four: "Pure Energy" for a definition of a joule.

Fig. 4.7 The 1946 Baker nuclear test at Bikini Atoll, which produced an explosion equivalent in force to detonating 21,000 tons of TNT. While several tens of kilograms of fissionable material was required to generate a runaway chain reaction, just a few grams of that material were all that was actually converted to the energy that powered the explosion. (Photo credit: U.S. Department of Defense)

body gained by digesting the first candy bar to throw the other candy bar with an equivalent kinetic energy, its initial speed would be just over 6.3 kilometers per second, or about 18 times the speed of sound. If, instead of eating the first candy bar, you converted its mass entirely into energy, the yield would be 5.3 million billion joules, just over 1 megaton[31], just under 1500 times the explosive yield of the Hiroshima bomb, or 19 million times the energy you received from digesting it (Fig. 4.7).

Conversion in the opposite direction—from energy to mass—does occur, but with far less dramatic results. It occurs in nuclear reactions and in some high-energy particle physics accelerator experiments. Scientists at Imperial College London are attempting to force interactions between high-energy X-rays and even higher-energy gamma rays to produce matter from energy[32].

> You may not feel outstandingly robust, but if you are an average-sized adult you will contain within your modest frame no less than 7×10^{18} joules of potential energy—enough to explode with the force of thirty very large hydrogen bombs, assuming you knew how to liberate it and really wished to make a point.
>
> Bill Bryson, *A Short History of Nearly Everything*

---

[31] In terms of explosive yield, a megaton is the equivalent of 1 million tons of TNT.
[32] UK discovery 'starts race' to turn light into matter: http://www.bbc.com/news/science-environment-27470034

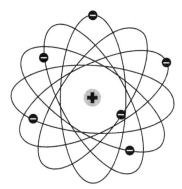 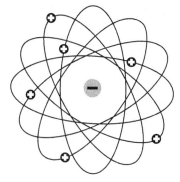

Fig. 4.8 Carbon (*left*) is composed of six negatively-charged electrons surrounding a positively charged nucleus made up of protons and neutrons. Its antimatter opposite, anti-carbon (*right*) has positively charged positrons orbiting a negatively charged nucleus made up of anti-protons and anti-neutrons

4.3 Does Antimatter Matter?

A fixture in science fiction, both television and film, is the concept that normal matter somehow has an "opposite" called antimatter. What may be surprising is that antimatter is not a science fiction construct, it is real, and not only does it participate in the process by which our Sun generates energy, doctors utilize a form of antimatter on a daily basis to diagnose disease.

If you have ever had a PET scan, your body has been probed by antimatter. "PET" stands for "positron emission tomography." Positrons are electrons that have a positive charge instead of the negative charge we are used to, and are a form of antimatter. Commander Data on *Star Trek: The Next Generation* (1987–1994) was said to have a positronic brain. That was actually a tip of the hat to Isaac Asimov's 1950 anthology of short stories, *I, Robot*, where robots had positronic brains—a detail that was kept in the 2004 film of the same name starring Will Smith.

The antimatter version of a proton is an anti-proton—particles with the mass of a proton, but with an opposite charge. In the classic, and fan-favorite, 1967 *Star Trek* episode "The Doomsday Machine," Mr. Spock's analysis of the planet killer's beam weaponry is "Pure anti-proton." What he means is that it is a beam of anti-hydrogen (Fig. 4.8)[33].

---

[33] There would, however, be no way for him to determine this by scanning the doomsday weapon as it fired its anti-proton beam; there would, if he scanned the target, and analyzed the gamma rays produced by annihilating protons and anti-protons.

4.4 Giving Matter Mass: The Higgs Boson

The Higgs boson had prominent guest-starring roles in the James Cameron film *Solaris* (2002), and in particle accelerators on the television series *Flash-Forward* and *Eureka*. Apart from Professor Proton[34] and Jimmy Neutron, how often do you recall a subatomic particle featured in such a high-profile Hollywood role?

The Higgs boson, sometimes referred to as the "God particle," is an elementary particle[35] that was first theorized to exist in 1964, and which scientists using the Large Hadron Collider at CERN discovered in 2012. The existence of the Higgs particle would confirm many existing scientific theories, but would also help explain why matter has the property of mass. (Technically, most of the mass of an atom arises from the huge amounts of energy involved in binding quarks together with the strong force, as per $E = mc^2$, but the quarks have some mass independent of that, which the Higgs is required to explain). Without mass there would be no gravity, and without gravity the Universe would be a very different place.

4.5 Flame or Fusion? Chemical and Nuclear Reactions

From a fundamental science standpoint, the film *Chain Reaction* (1996) starring Keanu Reeves, Morgan Freeman, and Rachel Weisz is… confusing. The premise is simple: Eddie Kasalivich (Reeves) and Dr. Lily Sinclair (Weisz) are University of Chicago researchers working to create a hydrogen-based source of clean, cheap alternative energy. When there is a huge explosion in their laboratory the very night following a successful test of the energy generation mechanism, and a fellow researcher is killed, Kasalivich and Sinclair are framed for murder and treason, and must go on the run.

On the surface, *Chain Reaction* is a typical action thriller with a seed of a science concept at its core and conspiratorial overtones. Taken at face value, the film is a fun romp. If science is important to you, and preventable basic science gaffes elicit "Oh please!" moments that forcibly eject you from the narrative, then you might want to give this one a miss.

How the hydrogen-based power source is described, and how its behavior is depicted, is self-contradictory throughout the film, and the vernacular between

---

[34] Bob Newhart's character in *The Big Bang Theory*.
[35] Meaning that, unlike a proton or neutron, it can not be further subdivided.

chemical burning and the nuclear "burning" of fusion[36] is interchanged early and often. That may not entirely be the screenwriters' fault. Scientists sometimes use the word *burning* both to describe the chemical combination of hydrogen with oxygen (oxidation), and the nuclear combination of hydrogen with hydrogen (fusion). In the beginning of the film, Dr. Alastair Barkley uses a flame, the chemical combination of hydrogen and oxygen, to describe his team's new energy generation method, later he says that they have achieved, "...sustained fusion for 4.6 milliseconds." The two processes, however, are quite far removed from one another: chemical burning occurs because of electron interactions, fusion occurs because of nuclear interactions, that occur in the absence of electrons—or, at least, electrons do not contribute.

One of the ways in which the Bohr model of the atom differs from reality is that planets can orbit at any distance from a central star. This is not true for atoms, where electrons can only exist in discrete, well-defined, orbits at fixed distances from the nucleus, and nowhere in between (as we'll see in Chapter Seven, "A Quantum of Weirdness," the picture is complicated somewhat by quantum mechanics, but it's a good way to think of things for now). Unlike planets, where there is typically only one planet per orbital path[37], different electron orbits can hold a fixed number of electrons, and that number differs depending upon the orbit.

Electron orbits of adjoining atoms do not, as a generality, overlap, or at least not much, because of a phenomenon known as the Pauli Exclusion Principle[38], which forbids two electrons in the same state from sharing the same orbit. Since the distances between nuclei and electron orbits are vast, and since electron orbits generally do not overlap, the overwhelming bulk of matter with which we come into contact on a daily basis ... is empty space. (More on this in the sidebar "Miniaturization: Black Holes vs. *The Fantastic Voyage.*")

> Atoms are mainly empty space. Matter is composed chiefly of nothing.
>
> Dr. Carl Sagan, Planetary Scientist, *Cosmos*

Atoms bond into molecules because most atoms act like they are obsessive-compulsive. They prefer to have their outermost orbital filled, and are willing a share some electrons to do it: some, like oxygen, prefer to accept electrons;

---

[36] For that matter, the movie title adds another level of ~~confusion~~ complexity, the term "chain reaction" is relevant to nuclear fission—unrelated to either chemical or fusion burning.

[37] Apart from Trojan Asteroids and other Lagrange Point co-orbiters, see Chapter Eleven: "Braver Newer Worlds."

[38] It is slightly more complex than that. More on the Pauli Exclusion Principle in the Chapter Six: "A Quantum of Weirdness."

some prefer to donate electrons. Atoms combine in an array of ways, to form an infinite amount of different compounds, simply because they want their outer orbital to be filled. This is chemistry.

For example, oxygen atoms are not only willing to accept two electrons to fill up their outer orbital, they "want" to do so very badly, so they are aggressive and will combine with most other atoms to do so. When oxygen combines with other elements, it's called *oxidation*[39]. When oxygen combines slowly with iron over time, we call that rusting. When it combines rapidly with wood or paper or natural gas to create a flame, we call that burning. So rusting and burning are the same thing. All that varies is the *rate* of oxidation.

Atoms that have their outer orbits already filled—gases like argon, neon, and xenon—are happy as is and, in general, unreactive with other elements, even oxygen. These are chemically inert, and are known as *noble gases*. Argon is used in sensitive areas, such as museum archives or server rooms, to extinguish fires by displacing all the oxygen in the atmosphere. Since this will knock anybody out who happens to be in the room when the fire suppression system goes off, it's occasionally portrayed on TV and film as a way for either good guys or bad guys to take out local security guards. Krypton is also in this category of unreactive elements. In the language of geology, "kryptonite" would be a mineral, perhaps even an ore, with a high concentration of krypton. Since krypton is one of the noble gases, it would form neither a mineral, nor an ore. Since it is chemically nonreactive, it is also one of the worst possible choices as Superman's "Achilles heel."

Hydrogen is an extremely explosive gas, and when hydrogen and oxygen gases combine in large amounts, they tend to do so explosively, releasing a lot of energy in the process—the lone byproduct of the reaction is H_2O, or water. So energetic is this reaction that liquid hydrogen is often combined with liquid oxygen as one form of rocket propellant. This explains why the airship Hindenberg was enveloped in flames so rapidly, because airships once used hydrogen gas to allow them to float. After the Hindenberg, and similar disasters, airship designers started using helium—like hydrogen, it is "lighter than air," but it is another of the noble gases, and does not burn.

Yet the energy released from the explosive combination of hydrogen and oxygen process pales in comparison to that generated by nuclear fusion. The core of our Sun is composed of extremely hot plasma—a random assortment of electrons and atomic nuclei. Since most of the Sun is composed of hydrogen plasma, most of those nuclei are simple protons. The fact that it is very hot in the core of the Sun means that all of these particles are moving very fast.

---

[39] Oxidation actually has a more general meaning, one that encompasses other elements and compounds beyond just oxygen, but we'll go with just this for now.

Electrons repel electrons and protons repel other protons by electrostatic repulsion—a fundamental manifestation of the electromagnetic force. In the high temperature of the Sun, however, protons can move so quickly that they can overcome this repulsion and approach one another close enough that the strong force kicks in, clicking together like a couple of Lego blocks. When the protons bond, that process releases a lot of energy[40]. This is the process known as nuclear fusion, the process that powers stars. It is a process that scientists have been trying to re-create in laboratories on Earth, in a way that produces more energy as output than it requires as input, since 1951.

Revisiting the film *Chain Reaction*, the energy generation process described in the film mixes terminology and visuals consistent with the chemical burning of hydrogen with terms appropriate to hydrogen fusion, and they do so about as well as oil mixes with water. At one point Dr. Shannon (Morgan Freeman) says that the mechanism uses a laser to initiate the energy generation process—which is consistent with conventional fusion research. At other times, we are told that acoustic waves, sound, is used in the process, which cause bubbles to collapse and emit ultraviolet light, a phenomenon known as *sonoluminescence*. While some have proposed that the temperatures in a sonoluminescing bubble are enough to permit fusion, there just isn't enough energy here to produce nuclear reactions.

Mistakes of this nature are often made in the script rewrite process, with far greater frequency than you might imagine, particularly in the rewrite process as it occurs in features as opposed to television. In television there is a showrunner, a high-ranking writer, to provide an overall consistency to scripts within a series. In feature films, writers are often brought in to rewrite previous writers, with little or no communications between them. Writer Zack Stentz elaborates, "You would never see in television what you have with some big blockbuster movies where you literally have different writers hired to write different versions of the same script, and then another writer comes in and stitches together the pieces that the studio and the director like the best into a Frankenstein's monster… You see the final film, and you see the little teeny traces and residues of, like, 'Oh wow, I bet three drafts ago this might have actually made sense'" (Fig. 4.9).

The film's narrative is also heavily reliant on the *new energy suppression conspiracy theory*, a conspiracy theory that dates back over a century, and recurs from time to time. The idea is that there is this guy somewhere who has created this radical new form of limitless energy, but it has been suppressed by the oil companies/auto industry/the government/the Illuminati because

---

[40] Actually, it releases a positron, a form of antimatter. The positron pair annihilates with a nearby electron yielding gamma radiation. Remember that matter, energy, and radiation are all inter-related.

Fig. 4.9 Keanu Reeves is Eddie Kasalivich in *Chain Reaction* (1996). (Copyright © Twentieth Century Fox. Image credit: Photofest)

profits would decrease/the economy would collapse/we're not ready for this technology.

It is fitting that our discussion of matter would finish on the topic of energy—they are different manifestations of the same thing. Burning fuels and fusion hydrogen are but two forms of generating energy, there are many others. We've stated that electrons orbit in various fixed orbits only, but these different orbits correspond to different orbital energies, and when electrons transition between orbits, they can emit electromagnetic radiation, a form of energy. When large unstable nuclei spontaneously split into smaller nuclei, they emit energy. To better understand how many of the other phenomena depicted in fictional science television and movies—from propulsion to weapons to the nature of the Universe itself—we are going to radiate to the next chapter, dude, and ask you to just bask in the positive energy.

Science Box: Go Ask Alice … All About Allometric Scaling

Go ask Alice,
When she's ten feet tall.

Grace Slick, Jefferson Airplane, *White Rabbit*

To the mouse and any smaller animal [gravity] presents practically no dangers. You can drop a mouse down a thousand-yard mine shaft; and, on arriving at the bottom, it gets a slight shock and walks away, provided that the ground is fairly soft. A rat is killed, a man is broken, a horse splashes.

J. B. S. Haldane, Evolutionary Biologist, "On Being the Right Size" (1926).

Giants have appeared in stories for as long as humans have been telling stories. In the many incarnations of *Alice in Wonderland*, from Lewis Carroll's 1865 novel to the various Hollywood interpretations spanning from 1933 to 2010, Alice grows in size significantly after eating cake, walking into an enchanted house, and eating magic mushrooms[41]. If her mass remains the same, then Alice's density would become so low that she would be in danger of blowing away in the wind. If, instead, Alice's mass grew proportionately with her height, then she becomes a fine example of *allometric scaling*.

Allometric scaling is a form of nonlinear scaling that determines how related values grow relative to an increase in linear size, and has the form:

$$y = kx^a.$$

One of the most common forms of allometric scaling is the *square-cube law*. First expressed by Galileo Galilei in his final book *Discourses and Mathematical Demonstrations Relating to Two New Sciences*[42] (known simply as *Two New Sciences*), the principle says that as an object grows in linear dimension, its volume grows at a greater rate than its surface area[43]. For a sphere, these two values are:

$$A = 4\pi r^2,$$

and,

$$V = \frac{4}{3}\pi r^3,$$

where r is the radius of the sphere, A is the surface area, and V is the volume. For increasingly larger objects, as one linear dimension (l) grows, area and volume grow:

$$A_2 = 4\pi \left(\frac{l_2}{l_1}\right)^2$$

$$V_2 = \frac{4}{3}\pi \left(\frac{l_2}{l_1}\right)^3$$

This law explains, for example, why larger planetary bodies[44] retain their heat longer than smaller ones. The surface area, the amount exposed to space through which heat can radiate away (more on this in Chapter Six: "Radiation: An All-Time Glow") increases as the square of the radius, but the volume, the three-dimensional area that encompasses the mass of the body, grows as the cube of the radius. Because a greater amount of mass holds a greater amount of heat (more on this in Chapter Five: "Pure Energy"), a larger body may have a greater ability to eliminate excess heat than a small one, but it has an even greater capacity to store it.

---

[41] She also starts growing again in the chapter "Who Stole the Tarts?" for no apparent reason.

[42] Or *Discorsi e dimostrazioni matematiche, intorno à due nuove scienze.*

[43] If we have two objects with different surface areas, the one with the greater surface area would require more paint to coat it. If two objects have different volumes, the one with the greater volume would require more water to fill it.

[44] This includes moons as well.

In a related way, this scaling law explains why large mammals like horses have a harder time staying cool than rodents. This also explains why gigantism is common among creatures that live in the very cold waters of both the Arctic and the Antarctic: larger creatures retain heat better.

Although people, animals, and kaiju are not spherical[45], they have been known to grow to fantastic sizes in Hollywood productions. Assuming a constant density, we might expect that the mass of these creatures would similarly follow the square-cube law. Density is an object's mass per unit volume, and is given by

$$\rho = \frac{m}{V},$$

where ρ is the density, and m the mass. Weight (with which we are already familiar from the previous sidebar on *John Carter*), is:

$$W = mg.$$

Solving the density equation for mass, and using that in the equation for weight, we have:

$$W = \rho g V.$$

If the density of a creature remains constant, then we would expect that an animal's (or a person's) mass would scale proportionately with its volume, or its length cubed. As the object grows increasingly larger, we have

$$W_2 = \rho g V_1 \left(\frac{l_2}{l_1}\right)^3 = W_1 \left(\frac{l_2}{l_1}\right)^3.$$

This is of the form $y = kx^a$, where $y = W_2$, $k = \rho g V_1 = W_1$, $a = 3$, and $x = l_2/l_1$. Using this equation, we can examine how the weight of a young lady like Alice might scale as her height grows.

Assume Alice, unaltered by cage or shrooms, is 5 feet 4 inches tall and weights 113 pounds, average height and weight for a slender woman from the United Kingdom (Alice *is* British, recall). When Alice is ten feet, or 120 inches, tall she weighs 745 pounds (339 kg). At fifteen feet all, the same height as the slender fish-faced alien Espheni on *Falling Skies*, she would weigh 2514 pounds (1143 kg). In *Attack of the 50 Foot Alice*, Alice would tip the scales at a whopping 93,000 pounds (42,300 kg = 42.3 metric tons, or 47 tons[46]).

In the original *Godzilla* (1954), the G-man was 50 meters tall. If Alice were to grow to that height, she would weigh in at 3.3 million pounds (1.5 million kg), or almost 1644 tons. Even though Godzilla is quite a bit beefier than Alice, it is unlikely that a creature that size would weigh in at the 20,000 tons he purportedly weighed[47], but he'd certainly be within a factor of ten.

In a January 18, 2014 interview in *Total Film* magazine, visual effects supervisor Jim Rygial said that the most recent incarnation of Godzilla would be the largest ever, standing 350 ft (107 meters) tall. If she was that all, Alice would weigh a colossal 1.1 billion pounds (almost 513 million kg), or 564,000 tons.

---

[45] Except in freshman physics problems.

[46] Any reference to "tons" implies "short tons", or 2000 lbs, as opposed to "metric tons" (which we will call out explicitly) which is the weight of 1000 kg of mass.

[47] From the Godzilla wiki: http://godzilla.wikia.com/wiki/Godzilla.

The cross-sectional areas of structural members like bones (or, with insects and crustaceans, exoskeletons) grow proportionately as the square of linear size, while mass grows at a greater rate. When he wrote *Two New Sciences*, Galileo showed that animal bones have to be proportionately larger for larger animals, anticipating (by almost 300 years) J. B. S. Haldane's 1926 essay in *Harper's Magazine*, "On Being the Right Size." Haldane said, "The most obvious differences between different animals are differences of size, but for some reason the zoologists have paid singularly little attention to them. In a large textbook of zoology before me I find no indication that the eagle is larger than the sparrow, or the hippopotamus bigger than the hare, though some grudging admissions are made in the case of the mouse and the whale. But yet it is easy to show that a hare could not be as large as a hippopotamus, or a whale as small as a herring."

Gigantic creatures whose form appears like actors in rubber suits would be in a state of disequilibrium, and would simply not have the structural strength to stand. Whether it was to address this issue explicitly or not, note that the 2014 incarnation of Godzilla is a lot more buff than his predecessors, but it's still probably not enough unless most of that extra girth is due to disproportionately larger bones than a much smaller reptile, or even a dinosaur. In fact, the primary reason why the largest animals that ever evolved on Earth are aquatic is because the buoyant force of water partially offsets the crush of gravity. Kaiju like Godzilla, and other behemoths like King Kong, star in some fun movies, but they are simply not plausible, and so must beg our suspension of disbelief.

When Bruce Banner transforms into the Hulk, there exists a sort of scientific trade-off. While the cross-sectional areas of his limbs grow at a rate beyond what would be dictated by allometric scaling in order to support the increased bulk, nevertheless, one has to ask, "How, exactly, does he get bulkier? From where does the extra mass come?" So when Banner Hulks out, his shape adapts to counter allometric-scaling-induced weakness, but he thumbs his big green nose at the conservation of mass. In Ang Lee's 2003 film *Hulk*, in a scene towards the end of the film, as Hulk reverts to Bruce Banner, perspicacious viewers will notice a large puddle of "Hulk juice" growing beneath him as he shrinks. So that film did make some concession to the Hulk/Bruce Banner mass imbalance.

Haldane said, "For every type of animal there is a most convenient size, and a large change in size inevitably carries with it a change of form." Nowhere is this statement more clearly reflected than when Dr. Bruce Banner transforms into the Hulk. Haldane's words underscore a more important point, though. Animals are the size they are for good reason. From a scientific standpoint, scaling them up or down simply does not work. It doesn't make sense, and such a process belongs more in the realm of fantasy, like *Alice in Wonderland*, than science fiction.

There is a denouement to this story. In Fig. 4B.1, we plot (red circles) the average weight of a slender woman from the UK versus height[48]. The best curve that fits these data[49] is:

$$W = (0.029)(H^{1.99}),$$

where W is weight in pounds, and H is height in inches. This also represents a form of allometric scaling, and is of the form $y = kx^a$, where $y = W$, $k = 0.029$, and $x = H$.

Plotting the height versus weight of an ensemble of people is a very different scenario than examining the weight of one individual as she grows. What this plot

---

[48] http://www.onaverage.co.uk/body-averages/38-average-female-weight.
[49] Determined by our plotting software.

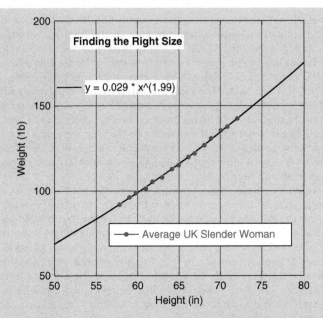

Fig. 4B.1 Alice's weight extrapolated to a height of 6.7 feet (80 in) (With a correlation (R) of 0.999. Statistics purists might prefer that R^2 is 0.999.)

shows is that, rather than mass scaling much faster than surface area, mass scales at
the exact same rate as surface area. This reflects the fact that animals in nature, or
at least slender women in the UK, do, in fact find their right size.

5

Pure Energy

If you want to find the secrets of the universe, think in terms of energy, frequency and vibration.
Nikola Tesla, Engineer/Inventor

The energy of the mind is the essence of life.
Aristotle

(On Loki in the Thor movies) The haters never win. I just think that's true about life, because negative energy always costs in the end.
Tom Hiddleston, Actor

It would be surprising if there were a town where the use of the term *energy* departs from its scientific meaning more than in Hollywood. A casting director may say to a colleague, "Oh I really liked his energy." In a coffee shop, you may hear a fellow discussing his last date, "She was amazing, I really resonated with her energy." Ask your friend why she chose not to move into the awesome apartment she just visited and you might get "I mean it was nice, but it had this really weird energy." Even in television we've heard characters like Mr. Spock refer to an alien life form as, "pure energy"[1,2].

People don't live in Los Angeles because we are tied to the same old, same old. We live in Los Angeles because of the intoxicating energy of new beginnings that permeate our city.

Marianne Williamson, Author

---

[1] Spock's "pure energy" quote from the 1967 *Star Trek* episode "Errand of Mercy" was sampled by the band *Information Society* and used in the 80s song "What's On Your Mind."
[2] The term *massive* is also misused, and perhaps more widely, than energy. One might say that the elephant at the zoo is "massive," and it is, but in colloquial usage "massive" means "really big" where, to a physicist, it means "possessing mass." A weather balloon might be bigger than a lump of lead, but the lead is more massive.

If none of those usages would be acceptable to a scientist—at least during work hours—then what *is* energy? The scientific definition is that energy is the ability to do *work*. Now "work" *is* a word in our sphere of experience … except that it, also, has a different, though related, meaning to a physicist. To most of us "work" means tasks that we do to earn a living or maintain our home or something that requires a large amount of effort, as in "you have to 'work' at relationships." Work, to a physicist, is the application of a force to move an object that has mass some distance.

If you push a stalled SUV through the snow to the nearest gas station, you have performed work, and probably quite a lot of it. If you push on a brick wall for hours on end until you collapse from utter exhaustion, you have done no work on the wall, since it has not moved[3].

Just as matter is conserved, there is also a Law of Conservation of Energy that says that energy can neither be created nor destroyed. We have already seen that energy and matter can be converted into one another. When physicists refer to energy, they refer to one of two different basic forms—kinetic energy and potential energy—and it is easy to show where these two forms of energy are converted from one to the other.

Any moving object that has mass has kinetic energy; an object at rest has zero kinetic energy. This ranges from the tiniest electron on up to spaceships to moons to planets to stars. The formula to calculate kinetic energy is:

$$E_{kinetic} = \frac{1}{2}mv^2.$$

That is, kinetic energy is equal to ½ of the product of the object's mass and the square of its velocity. The unit of energy most commonly used by scientists is a *joule*[4]. Roll a 1-pound (453 gram) hand weight across the mat at the gym at a rate just slightly faster than a person typically walks, and that weight will have 1 J of kinetic energy. At a typical walking speed of 3 mph (about 1.34 m/s), a 100 kg person[5] has about 90 joules of kinetic energy; at a four-minute mile pace (15 mph or 6.7 m/s), that rises to 224 joules. A 0.2-g bullet fired from an AR-15 rifle has almost four times that amount, or 981 joules of kinetic energy. An 820-kg Smart Car traveling at 25 mph (40 km/h) has just over 51,000 joules of energy. If the containment vessel in *Angels and Demons* held one gram of anti-matter, the energy released in the event of a containment vessel failure would be

---

[3] You will have done some work within the internal structure of your muscles, which is why you're tired. You have expended energy internally, but you have done zero work on the wall.

[4] Many other energy units exists, including the BTUs with which Morpheus measures the human body's heat output in 1999s *The Matrix*. Somewhat unglamorously, BTU stands for "British Thermal Unit." One BTU equals 1055 joules.

[5] Both SAC and KRG qualify.

1.8×10^{14} joules (by comparison, the yield of the Hiroshima bomb was about 6×10^{13} joules). Each second, our Sun produces 3.8×10^{26} joules of energy[6].

Science Box: Armageddon and the Cretaceous-Paleogene Event

On the surface, the 1998 film *Armageddon* had a lot going for it. It had a large budget, it was a science fiction action adventure in which brave heroes race to prevent the end of the world, it had Bruce Willis. Still, *Armageddon* is a movie that many scientists love to hate. The science in the movie is heinously bad, and examples of appalling science occur early and often. Just 34 seconds into the movie (give or take), Charlton Heston claims in a voiceover that the asteroid impact that led to the extinction of the dinosaurs[7] "… hit with the force of 10,000 nuclear weapons." One should replace "force" in that sentence with "energy," since it makes more sense, but this oversight in nomenclature is a small goof compared to a myriad of others in the film. What about the claim, though, that the impact was the equivalent of 10,000 nuclear weapons? How reasonable is that? It's a very simple "back of the envelope" calculation to find out.

Let's give Mr. Heston the benefit of the doubt, and assume that all of the kinetic energy of the asteroid impactor is converted into explosive yield. The equation for kinetic energy is

$$E_{kinetic} = \frac{1}{2} m_{impactor} v^2.$$

What is the mass of the object? We can estimate the mass if we know the volume and the density of the object. Here is a prime example of why this is known as a "back of the envelope" calculation: there are many unknowns here, but by making a few reasonable assumptions, the result can prove incredibly insightful. We also don't know the shape of the object that impacted Earth, and its volume is dependent upon its shape. What we do know, from seismic reflection studies of the 65-million-year-old crater that remains in Central America, is the size of the impact structure, and from laboratory experiments we can infer from this that the diameter of the impactor was about 10 miles, or 16 km, across. So let's assume that the object was spherical—it may have been highly irregular, but with no evidence spherical is as good a guess as any. The volume of a sphere is:

$$V = \frac{4}{3} \pi r^3.$$

If the diameter is 16 km, then the radius is 8 km, and plugging that into the volume equation, we wind up with a volume of 2145 km³ or 2.14×10^{12} m³.

We also don't know the density of the object. Some asteroids are composed of rock, some metal, some rock and metal. For now let's assume that the asteroid was rocky—which is the lowest density of the three types—and so has a density of roughly 3000 kg/m³ (metal would be approximately 7500 kg/m³, and metal/rock somewhere in between). The equation for density (ρ) is:

$$\rho = \frac{m}{V}.$$

---

[6] For a longer list of examples, check the Wikipedia article on the *Joule*, with examples ranging from one nanojoule to yottajoule (one septillion joules).
[7] Formerly called the Cretaceous-Tertiary (K-T) Event, now called the Cretaceous-Paleogene (K-Pg) Event.

Rearranging to solve for mass, we have:

$$m = \rho V$$

Inserting our values for density and volume, the mass of the asteroid was roughly 6.4×10^{15} kg.

The only thing remaining to solve for kinetic energy is a reasonable value of velocity. Minimum asteroid impact velocities are in the 11 km/s range. Typical velocities would be close to 17 km/s, but for now let's use the lower, more conservative, value[8]. Inserting everything into our equation for kinetic energy yields:

$$E_k = \frac{1}{2} m_{impactor} v^2 = \frac{1}{2}\left(6.4 \times 10^{15}\,\text{kg}\right)\left(11,000\,\text{m}\big/\text{s}\right)^2 = 7.74 \times 10^{23}\,\text{J}$$

That is 774 zettajoules (where "zetta" means 10^{21}).

Comets have densities much lower than asteroids, in the 1000–2000 kg/m³ range, but they have higher impact velocities, and the equation for kinetic energy shows that the energy is highly dependent on the velocity of the object, because the velocity is squared. So let's do the calculation for a comet as well. Some comets can be very porous, so we will use the lower end of the density range, which gives us a mass for our 16-km-diameter comet of 2.1 x 1015 kg. Comet orbits tend to be very oblong, so if its most distant point from the Sun is in the deep outer Solar System, and its closest point is at Earth, the impact velocity is more like 12 km/s[9]. Solving:

$$E_k = \frac{1}{2} m_{impactor} v^2 = \frac{1}{2}\left(2.1 \times 10^{15}\,\text{kg}\right)\left(12,000\,\text{m}\big/\text{s}\right)^2 = 1.54 \times 10^{23}\,\text{J}$$

That is 154 zettajoules.

A 1-megaton nuclear weapon, which is on the large side, liberates 4.18×10^{15} joules of energy, so our values for our asteroid and comet impactors are between 36 million and 185 million times higher. That means that a reasonable value for the number of nuclear weapons to equal the explosive yield of the K-Pg impact is between 3600 and 18,500 times beyond the number of weapons stated in the film.

In reality, scientists have calculated, using much more sophisticated methods, that the K-Pg impact event liberated approximately 420 zettajoules of energy, which is nestled between the two values we calculated, so two things become immediately clear. First, we see that the back of the envelope calculation can be a very useful tool. More importantly, had *Armageddon* had a science consultant, the opening lines would have had far greater dramatic … impact.

---

[8] Asteroid impact velocity from the Earth Impacts Effect Program, a joint project by Imperial College, London and Purdue University: http://impact.ese.ic.ac.uk/ImpactEffects/. Warning: the output of this program can be very sobering.

[9] For those keen on the technical details, this value was calculated by solving the vis-viva equation for an object on an orbit with an apocentric distance of 50 AU, and a pericentric distance of 1 AU.

Potential energy can be viewed as the "potential to have kinetic energy." For an object moved over short vertical distances, immersed in Earth's gravitational field, potential energy is simply:

$$E_{\text{potential}} = mgh.$$

That is the mass of the object times the acceleration of gravity (which we derived in the last chapter) times the height you lift the object. So at the gym, lift a one-pound hand weight a little more than 8 ½ inches (22 cm) off the ground, drop it, and it will have one joule of kinetic energy when it hits the ground. You gave the hand weight potential energy when you lifted it, and that energy was converted to kinetic form as it accelerated to the ground. Of course the weight needed you to give it kinetic energy in order to lift it and gain that potential energy in the first place, so we see how these two forms of energy can be converted from one to the other.

The exchange between potential and kinetic energy is a ceaseless process, going on everywhere around you. Examples, both on screen and in the real world, are easy to find. To understand better how potential energy and kinetic energy can be interchanged, imagine Spider-Man perched on the window ledge of a skyscraper[10]. In any case it is a long way to the ground, so you would be correct in assuming that Spider-Man has a lot of potential energy. A nearby construction crane is perched over the street in front of the building on which Spider-Man is currently crouched. He shoots a web that sticks to the crane, and leaps off his ledge. Spider-Man will accelerate towards the ground, potential energy being converted into kinetic, as he moves ever faster. As the tension of his web pulls his trajectory into an arc, Spider-Man will reach a maximum vertical distance beneath the crane, but a minimum distance from the street below. As his trajectory arcs upwards, kinetic energy is turned back into potential energy—as Spider-Man ascends higher, his speed is reduced, until, at the same level as the ledge where he started, his speed drops to zero (Figs. 5.1, 5.2).

---

[10] You can pick your own screen adaptation for reference here. SAC prefers either 2004s *Spider-Man 2* or the 1978–1979 TV series *The Amazing Spider-Man;* KRG just likes Spider-Man and will take what he can get.

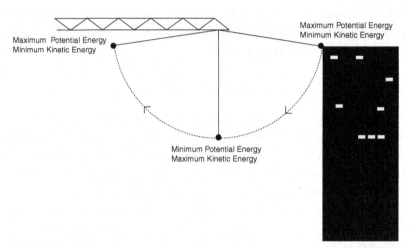

Fig. 5.1 When Spider-Man leaps off a ledge attached to a rope made of webbing, his potential energy is converted to kinetic energy and back again as he first accelerates, then slows to a halt at the top of his swing. Here, Spider-Man is illustrated as a ball for copyright reasons, and because physicists really like assuming objects are spherical.

Fig. 5.2 A nonspherical approximation of Spider-Man. Spider-Man's super power revolves around his ability to covert potential energy into kinetic energy and back again. Copyright © Columbia Pictures, Marvel Studios. Image credit: MovieStillsDB.com

Science Box: The Fall of Total Recall

A particularly extreme example of energy transformation is front and center in the 2012 remake of *Total Recall*[11]. *Total Recall* was a fun romp, but from a science perspective the film suffers early and often. Eschewing the Martian setting of the 1990 film (and the 1966 book *We Can Remember It for You Wholesale*)[12], the 2012 remake is set on an Earth rendered largely uninhabitable by chemical warfare, with only two regions capable of supporting life: The United Federation of Britain, and The Colony (Australia). A long shaft through the interior of Earth connects London to The Colony, and travel back and forth between the two locations requires passengers to step into a skyscraper-sized elevator car called The Fall—except this elevator car has no cables.

The Fall is simply an extreme example of a method of transportation called a *gravity train*, first suggested by British scientist Robert Hooke to Isaac Newton in the 17th century. To travel from one terminus to the other, The Fall is simply dropped down the shaft—the elevator car accelerates, turning potential energy into kinetic energy as gravity draws it towards the center of Earth. After passing the center of Earth, the car is travelling at high velocity, but now gravity—always pulling towards the center of Earth—begins to brake the car, slowing it, and converting its kinetic energy to potential energy. Because the distance to London and the core, and the core to Australia, is the same, the car runs out of kinetic energy at the exact moment it arrives at the surface. For a moment it comes to a complete halt, whereupon clamps grab the elevator car before it can begin falling back towards the center of the Earth.

It sounds simple, but in reality, the physics behind The Fall is actually fairly complex. In the film the car was depicted as accelerating at 1 G during the descent, and decelerating at 1 G during the ascent, and people able to walk about except at the midpoint of the trip. At that point all passengers were briefly weightless, and the cabin performed "the flip," where the seats were automatically reoriented for the ascent. If the descent was purely vertical, the passengers actually would be weightless, in free-fall, the entire time (a diagram of the shaft is presented briefly on screen that depicts it curving slightly to avoid the Earth's inner core, but not enough to affect our point). So there is no good reason for the "flip" in the orientation of the passenger seats at the midpoint of the trip.

In order to examine other aspects of the system even cursorily, we will make some (vastly) simplifying assumptions. In the Chapter Four Science Box *John Carter's* Flee Circus" we derived the acceleration due to Earth's gravity, which is directly proportional to Earth's mass:

$$g_{Earth} = \left(-\frac{GM_{Earth}}{r^2} \right).$$

---

[11] We don't believe this was the filmmakers' intention, but we rooted for Kate Beckinsale's dedicated, skilled, and gutsy undercover agent character the entire movie.

[12] The change of venue may have been to distance the film from (relatively) recent box office disasters like *Mission to Mars, Red Planet*, even *John Carter*: http://www.slate.com/articles/arts/the_middlebrow/2012/03/john_carter_whatever_happened_to_the_mars_of_h_g_wells_and_ray_bradbury_.html

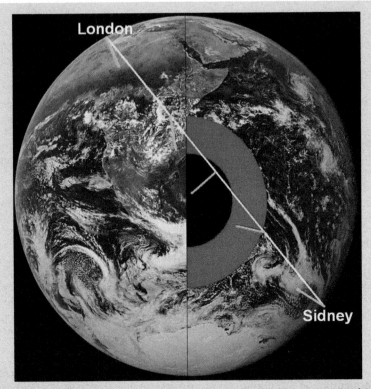

Fig. 5A.1 The Fall from the 2012 remake of *Total Recall*. Shown are Earth's solid inner core (*black*) and molten outer core (*red*). The *yellow line* shows the approximate path through Earth—from London to Australia—and the *green vectors* show the direction of "down" at various points on the trip. Photo courtesy of NASA-Caltech-JPL.

In that equation, M_{Earth} is the mass of Earth, but it is the mass of Earth at a radius less than your current position. At the surface, the mass is all of Earth, but if you were descending in The Fall, the value of g would be ever-changing as the distance to the center of Earth, and the mass between the car and the center, is ever-changing.

Figure 5A.1 shows the path of The Fall through Earth. London and Sidney are not *antipodes* of one another—they are not on exact opposite points on the globe. So a chord connecting London and Sidney (the point in Australia closest to being antipodal to London) does not run through the center of Earth, and the length of the chord connecting the two cities is about 10,550 km, or 84 % the diameter of Earth (13,462 km). The travel time quoted in the film is 17 min (0.28 hours), meaning the average speed during one trip would be:

$$v = \frac{d}{t} = \frac{10550 \text{ km}}{0.28 \text{ hr}} = 37235 \text{ km}/_{\text{hr}},$$

which is 23,137 mph, or approximately 30 times the speed of sound[13].

Although we know that the acceleration would be time-varying, let's look at how long the trip would take under the constant 1 G acceleration depicted in the film. The distance an object travels while falling is:

$$d = d_0 + v_0 t + \frac{1}{2} g t^2,$$

where d is the distance, d0 the starting distance (since The Fall starts at the surface of Earth, we can assume this is 0), v0 the starting velocity (also 0), g the acceleration (9.81 m/s2), and t the time in seconds. Solving for *t*:

$$t = \sqrt{\frac{2d}{g}} = \sqrt{\frac{2(5.277 \times 10^6 \,\text{m})}{9.81 \,\text{m}/\text{s}^2}} = 1037 \text{ s} = 17.28 \text{ min},$$

which is the midpoint travel time, so doubling that, the entire travel time is 34 ½ minutes, and the average speed "only" 18,316 km/h (11,381 mph) or 15 times the speed of sound. Since g decreases as the car approaches the core, the travel time would be still longer.

It turns out that for a gravity train, the lone factors that determine the travel time between any two points on Earth's surface are the gravitational constant G and the density of Earth. If Earth were a constant density[14], the travel time between any two points would be a constant 42 min (for an average speed of "only" 12 times the speed of sound).

There are two more complexities that conspire against the car even reaching this speed, both involving friction. Friction is the enemy of mechanical efficiency, and all machines—from the simplest to most complex—suffer its effects. Friction, which occurs when two objects rub against one another, is like an energy vampire, turning energy that might be used to do useful work into heat dissipated either within the system or to the environment.

The descent was not vertical, though. In Fig. 5A.1, the green vectors point "down" (the direction of the local gravitation vector) for various points along the journey. Since the shaft does not run vertically there is always a component of gravitation forcing the car up against the shaft wall, and the passengers most definitely would have weight at the mid-point. Friction with the wall would dramatically slow both the descent and ascent, and limit the top speed.

What if the car could be held away from the shaft wall by *magnetic levitation*, or *maglev*, the same technique used to levitate high-speed trains in many parts of the world? Friction would *still* be the enemy: air friction. Since characters are shown outside of the car during the trip without respirators, air is not evacuated from the shaft while the car is traversing The Fall[15]. There is no known material that could

---

[13] Many factors affect the speed of sound, including elevation, humidity, even atmospheric composition.
[14] Average density of Earth's core is about 10 g/cm³; average density of the rocky envelope is about 4 g/cm³.
[15] Since the shaft of The Fall is not evacuated, either air is allowed to flow along the sides of the elevator car (creating more friction) or a 10,550 column of air is pushed out of the arrival terminus of The Fall over the duration of one trip. Then there is the vacuum created as an equal portion of air is sucked in at the departure terminus. These would not be safe places to be near, yet they are shown at the heart of urban centers.

withstand the temperatures generated by the air friction associated with an object travelling at 15, let alone 30, times the speed of sound in the atmosphere. People outside of the car, subjected to those air speeds, would simultaneously be torn apart and vaporized (even pilots ejected from fighter craft travelling at mach, or even high sub-mach, speeds risk having limbs torn off by air friction[16]).

In Chapter One, Jon Amiel spoke of conceits in science fiction storytelling—the "gimme" that you need the audience to make to tell the story. Talking specifically about his film *The Core* (2003), lambasted by many as having some of the worse science ever depicted onscreen, Amiel shares, "In The Core our big gimme to the audiences was, 'Can you believe that a material exists that would withstand the heat and internal pressures of the Earth's core?' It was a big gimme. I think probably the reason the movie didn't do as well as it should have done is a lot of people just didn't buy that such a thing … they just wouldn't make that leap. Sometimes, as I said, you need to make that fictional leap in order to illustrate some bigger, broader truth."

Under the heading of "Journalistic Fairness," in addition to asking the audience to make the leap that vivid memories can be implanted into a person's brain[17], The Fall in the 2012 incarnation of *Total Recall* requires the audience to buy into most *every single conceit* of *The Core*, and far more. Logically, then, irrespective of how much fun the film was to watch, shouldn't there be an outcry resulting from the science in *Total Recall* at least equal in magnitude to that of *The Core*? Not only was *Total Recall* not equally lambasted for The Fall, some film critics even lauded its use[18]. Clearly context, perhaps even pacing, is another complicating factor influencing the science that viewers will, and will not, accept.

5.1 Feeling the Heat

Two terms that, in colloquial usage, are often used interchangeably are "temperature" and "heat." "If you can't stand the heat, get out of the kitchen" isn't bad, but "If you can't stand the temperature, get out of the kitchen" is probably more accurate. We say that we are having a "heat wave" when we mean "we are having a stretch of high temperatures." Temperature is the measure of the average kinetic energy of the atoms/molecules of a substance. In essence, it is a measure of how fast the atoms are moving on average. Increase the temperature, things move faster.

There are three temperature scales used commonly today. Most of the world uses the *Celsius scale*[19] based upon the properties of water—the freezing point of water is 0 degrees and it boils at 100 degrees.

---

[16] In the pilot episode, the writers of CW's *The Flash* carefully noted how Barry Allen's suit protects him from harm due to air friction when he runs at super high speeds.

[17] Which may not be a huge leap, but that's a topic for another time.

[18] "Total Recall: Not a Total waste of time" by Dana Stevens: http://www.slate.com/articles/arts/movies/2012/08/the_total_recall_remake_reviewed_.html

[19] Which is very close to the Centigrade scale. There is an ever-so-slight difference, but not enough to care about for our purposes here.

The *Fahrenheit scale*, while considered out of date for most of the world, is still used in several countries today, including the United States. On this scale, water freezes at 32 degrees while it boils at 212 degrees. While most of the world thinks the countries still using Fahrenheit are backwards and should "make the leap to the 20th century," its few proponents argue that it is a temperature scale that is most aligned with our daily experience. On the Fahrenheit scale, if it is 0 degrees outside it is "really chilly," while if it is 100 degrees, it is "quite toasty." On the Celsius scale, 0 degrees means "fairly chilly," while 100 degrees outside means "we're all dead."

Another scale used by scientists is the *Kelvin*, or absolute, scale. The temperature difference between two values on the Kelvin scale is the exact same as with the Celsius scale, but the starting points are different[20]. (If you listen carefully to the dialogue in 1984s *Dune*, you'll hear desert temperatures given in Kelvin.) The zero point of the Kelvin scale is known as absolute zero, the coldest temperature possible, and it corresponds to -273.15 degrees Celsius. Contrary to popular belief, this is not the temperature at which atomic motion ceases. If an atom were to stop, both its position and velocity would be knowable. This violates a fundamental rule of quantum mechanics called the *Heisenberg Uncertainty Principle*, which we will discuss more in Chapter Seven, "A Quantum of Weirdness." Absolute zero can be thought of as the point where all possible energy has been extracted from a substance.

There is another quirk of the Kelvin scale. While it is proper to give a reading in "degrees Celsius" or "degrees Fahrenheit," one properly gives temperature in "Kelvins." So water freezes at 273.15 Kelvins, and it boils at 373.15 Kelvins. This is a mistake that even many scientists make—even those who know better[21]. Pay close attention in *Star Trek: The Next Generation*, and notice that at the beginning of the series, characters said "degrees Kelvin," and in the later seasons, they said simply "Kelvins." Clearly somebody corrected the writers. Also, in the second season episode "The Royale," Engineer Geordi LaForge reports that the temperature on the planet the *Enterprise* is orbiting is "minus 291 degrees Celsius." This corresponds to a temperature below absolute zero and is a much bigger "Oops" than saying "degrees Kelvin."

> Ah, Kirk, my old friend, do you know the Klingon proverb that tells us revenge is a dish that is best served cold? It is very cold in space.
>
> Khan Noonian Singh, *Star Trek II: The Wrath of Khan*

---

[20] There is also an absolute scale using Fahrenheit graduations called the *Rankine scale*. Given so few people, globally speaking, use Fahrenheit—and none of them are scientists—do you think anybody uses Rankine?

[21] Seriously, though, on the scientific "Oops" scale, saying "degrees Kelvin" is not a biggie and few will scold you.

To a physicist, *heat* is not synonymous with temperature. Heat is the transfer of energy from one object to another object, or from an object to its surroundings, that arises as a result of temperature differences. It is a measure of energy flow—just like wind flows from regions of high pressure to regions of low pressure, energy flows from places where there is a lot, to places where there is less. So, counterintuitively, it is wrong to say that an object has heat independent of any energy flow.

There are three methods of heat transfer: conduction, convection, and radiation. Conduction is the heat transfer mechanism with which we are most familiar. Energy is transferred from objects at higher temperatures to objects at lower temperatures when they are in contact. Touch a hot plate, energy is transferred from the plate to your cooler hand, and, "Ouch!" When you step into a bath, anything below skin temperature (about 91 degrees Fahrenheit or 33 degrees Celsius) feels cool, and your body is transferring heat to the bath. Convection involves vertical circulation and temperature equalization that occurs only when a fluid is heated from beneath, such as a pot of water on the stove, a waterbed mattress with a heating element underneath, or deep within planetary and stellar interiors. Convection deep within Earth's *aesthenosphere* causes the tectonic plates on the surface to shift, resulting in earthquakes, while convection in the Earth's outer core of liquid iron produces circulating electrical currents that generate Earth's magnetic field.

An interesting consequence of temperature is that all objects above absolute zero emit electromagnetic radiation known as *blackbody radiation*. If an object is hotter than its environment, then it will radiate energy into the environment[22]. If the environment is hotter than the object, then the object will warm as it absorbs ambient radiation. This explains why a planet, hot at birth, eventually cools down—it radiates its heat to space. (More on this in Chapter Six: "Radiation: An All-time Glow.")

Temperature is an absolute measure of the energy stored within a portion of matter, yet different objects can have the same temperature while possessing very different amounts of *total* energy. For example, the tungsten filament in a halogen lamp typically reaches temperatures of about 3300 Kelvins as the tungsten atoms are excited by electricity, increasing their kinetic energy to the point where they emit white light. For comparison, the temperature of a blast furnace is only about 1500 Kelvins. The reason why incandescent lights don't vaporize everything in their immediate surroundings is because their filaments are small, so they contain a lot less total energy than several tons

---

[22] Kudos to the designers of the *Red Dwarf*. There are clearly heat sinks—passive emitters designed to radiate excess heat to space—on her main engine's thruster bell.

of molten metal, which allows the lamp filaments to be thermally insulated relatively easily.

Conversely, the average temperature of the water in the Earth's oceans is about 276 Kelvins, much chillier than your morning coffee, but the oceans are so vast that even relatively small water temperature fluctuations can cause significant heating and cooling effects on the atmosphere.

Physicists call a large massive object that can store a lot of thermal energy, one that can absorb or donate a large amount of energy but whose temperature does not change appreciably in the process, a *heat reservoir*. A great example where an understanding of this concept would be useful would be on a television show like *Survivor*[23]. Typically on the first night, contestants are concerned with getting a roof over their heads, rather than thermally insulating themselves from the ground. The ground is an excellent example of a heat reservoir. Even a 10 degree temperature difference between the ground and a contestant sleeping directly upon it without some form of insulation can lead to dangerous heat loss and hypothermia.

5.2 Energy Generation: Synthetic Joules

What powers our laboratories, our hospitals, our starships, our time machines? Energy production is a hot topic in society today, and how energy is generated is often a key dramatic component in fictional science productions. In both the real world and fictional narratives, energy is generated by chemical, nuclear, and quantum processes, and that energy is converted from its natural form into another form (often electrical energy) that can be more easily used by technological devices.

One of the more common energy generation techniques, both onscreen and in real life, involves liberating the chemical potential energy of the hydrogen locked within the molecules of hydrocarbons—typically fossil fuels. Combustion engines like those in today's vehicles release the chemical potential bound within hydrocarbons like natural gas or methane (CH_4), octane (C_8H_{18}), hexadecane or diesel ($C_{16}H_{34}$), ethanol (C_2H_6O), or some combination thereof, by combining them with oxygen. The reaction explosively coverts the chemical potential energy into the kinetic energy used to push an engine's pistons. Batteries unlock the chemical potential of reactants stored within to generate electricity in a different way that, unlike combusting a fuel

---

[23] KRG actually applied to be on *Survivor*, and is the only person in the history of the show to have Richard Hatch, star of both old and new *Battlestar Galactica*—not the nudist/winner from season one—in his application video.

with oxygen, can be easily reversed in many cases, hence rechargeable batteries. Unfortunately, the chemistry of batteries has proven very difficult to significantly improve, which is why you have to charge your smartphone at least once a day, and why the threat of running out of juice to power combat suits was a recurring theme for the protagonists of the 2014 movie *Edge of Tomorrow.*

Fuel cells take a fuel like hydrogen or methane and, through a much slower and controlled chemical reaction than in an internal combustion engine, generate electricity. Fuel cells powered the Gemini and Apollo missions of the 1960s and 1970s, as well as the space station orbiting the planet Solaris in *Solaris* (2002). In the Apollo missions the simple hydrogen/oxygen fuel cells not only generated electricity, but, as a byproduct, produced drinking water for the astronauts. On the Apollo 13 mission, however, a spark within a fuel cell ignited an oxygen fire. The resulting rapid combustion produced an explosion that crippled the spacecraft and almost killed the crew, prompting the most famous distress call in history: "Houston, we've had a problem." (A reasonable depiction of events is presented in the 1995 movie *Apollo 13* but not without some dramatic license. For example, the grounded astronaut Ken Mattingly did not actually help work out how to get enough power to the command module for a safe landing[24].)

In today's power plants, either the chemical potential energy locked away in coal is released through burning, or the energy locked within unstable radioactive nuclei is used to heat some intermediate fluid (typically water, but molten sodium metal is used in some advanced designs) and spin a turbine. Attached to the turbine is a magnet that spins within a coil of wire, which is what actually generates electricity. Should fusion power become practical it may well become the method of choice for generating electricity, liberating the potential energy found within the nuclei of atoms (Fig 5.3).

Solar cells or, more correctly, *photovoltaic cells*, are composed of semiconductor materials that, when exposed to light, generate electricity directly without having to go through intermediate processes like spinning a turbine to turn a generator. This photovoltaic effect is related to the photoelectric effect—first described by Albert Einstein[25]—where certain metals become electrically charged when exposed to light of a certain minimum energy. Solar

---

[24] If you're interested in an account that doesn't take any dramatic license and which focuses on the controllers working to save the mission, check out SAC's article "Apollo 13, We Have A Solution," on *IEEE Spectrum*'s web site.

[25] It is for his description of the photoelectric effect that Einstein won his Nobel Prize, not for Relativity. Einstein won the 1921 Nobel Prize for Physics retroactively in 1922—the same year that Neils Bohr won the Nobel for his understanding of the structure of the atom, and how they emit radiation.

Fig. 5.3 MIT's experimental fusion reactor. Buried within a cocoon of pipes and monitoring equipment, the multistory cylinder at its heart contains a toroidal magnetic bottle, called a *tokamak*, for heating and containing a plasma with temperatures so high that fusion occurs between atoms. Unfortunately, the reactor can only operate for short periods and requires more energy to run than can be extracted. A researcher can be seen standing on top of the tokamak for scale. Photo: Stephen Cass.

cells are becoming increasingly commonplace today on all sorts of buildings (although they still only account for a fraction of a percent of our energy needs), but they have provided electrical power to spacecraft in both real life and in the cinema since the dawn of the space age (Fig. 5.4).

Movies however, lay claim to more exotic ways of generating energy that, at least for now, still fall within the purview of science fiction. Most of us have watched enough science fiction to know that when matter and antimatter come into contact, there is a colossal explosion: an Earth-shattering *Ka-Boom!* This phenomenon was also at the heart of the mainstream 2009 thriller *Angels and Demons* starring Tom Hanks. An Earth-shattering Ka-Boom occurs only

Fig. 5.4 The original starship *Enterprise*, and all her fictional descendants, used a matter-antimatter reaction to generate power. The reaction is so potentially dangerous that a mechanism to jettison the vessel where the reaction occurs is a standard feature. Copyright © Paramount Television. Image credit: PhotoFest.

when large amounts of matter and antimatter come into contact, but irrespective of the amounts, when a particle and its antiparticle collide, the entirety of their mass is converted into energy (with the amount given by $E = mc^2$) in the form of gamma rays in a process called *pair annihilation.*

We have already seen that when matter is converted into energy, a little goes a long way. You do not have to be a die-hard Trekker to recall that the various incarnations of the Starship *Enterprise*[26] are powered by matter/antimatter reactions[27], but a variation of the pair annihilation method of generating energy is to tap the energy of the vacuum itself, used in the zero-point modules in the *Stargate* franchise (vacuum energy is explained in Chapter Seven, "A Quantum of Weirdness.") How pair annihilation is harnessed to generate usable power is never explored in detail on *Star Trek* itself, but we can assume that the *Enterprise* does not rely on energy generated from a matter/antimatter reaction to drive a steam turbine (Fig. 5.4).

Another word whose technical definition can cause problems is *power*. While the word "power" is often used synonymously with "energy," strictly speaking, power is a measure of energy expenditure over time. The unit of power, the Watt, is one Joule per second. How high you lift a weight at the gym requires a certain amount of energy, while power measures how quickly you lifted it.

---

[26] As well as all Starfleet Wessels.
[27] Moderated, of course, by dilithium crystals. Now those *are* fictitious.

Still, even though Hollywood may have co-opted the word "energy" for many different purposes in real life, and may be somewhat nebulous about its proper usage in TV/film, there is another word that, because of angst it induces in viewers by its mere mention, really wins the word abuse race and which is the subject of the next chapter: *radiation*.

Science Box: Pushing the Moon Out of Orbit

In "Breakaway", the pilot episode of the 1970s British television series *Space: 1999*, nuclear explosions on the far side of the Moon—created when all the radioactive waste Mankind has been storing there for decades reaches a critical mass—push the Moon out of Earth orbit. The series then follows the lives of the 311 men and women trapped on Moonbase Alpha as the rogue Moon wanders our Galaxy.

Is this scenario the least bit plausible? Assuming that nuclear waste could explode—which is highly improbable in the first place since most nuclear waste is not bomb-grade material—could nuclear explosions push the Moon out of orbit? It all boils down to the concept of energy. An object's total energy is the sum of its potential energy and kinetic energy:

$$E_{total} = E_{kinetic} + E_{potential}.$$

For the Moon, the kinetic energy would be:

$$E_{kinetic} = \frac{1}{2} m_{Moon} v^2.$$

The Moon is, on average, 384,399 km from Earth. Its orbit is very nearly circular, so let's make the simplification that it is, therefore its speed, v, is constant. If we assume that, and its period is 27.3 days, then the Moon is speeding along at 1024 m/s. Its mass is 7.36×10^{22} kg. Plugging those values into the above equation, its kinetic energy is 3.88×10^{28} joules.

In the chapter we had one equation for potential energy for objects moved short vertical distances within a gravitational field. For celestial objects at vast distances, the equation is different[28]. The equation for the Moon's gravitational potential energy relative to Earth is (notice that, although the equations are similar, this is a $1/r$ relationship, not $1/r^2$ as with the gravitational force):

$$E_{potential} = -\frac{GM_{Earth} m_{Moon}}{r}$$

Where G is the Universal Gravitational Constant, which we have already seen. The mass of Earth is 5.97×10^{27} kg, giving us a gravitational potential energy of -7.63×10^{28} joules. If we add its potential and kinetic energy, Earth's moon has a total orbital energy of -3.77×10^{28} joules.

This value of energy is negative because an object in a bound trajectory—an elliptical or circular orbit—always has negative total energy. If a celestial body or satellite has a trajectory with a positive total energy relative to another, in our case the Moon relative to Earth, it is *unbound* and is travelling too fast to remain in orbit or to be captured into orbit in the first place. So what we have to determine is

---

[28] Actually, they're pretty much the same, but that takes more math than you want to see in this book.

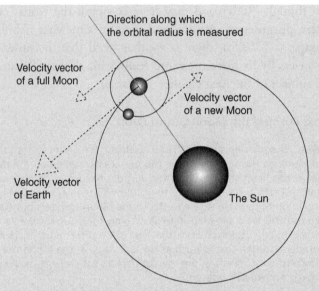

Direction along which
the orbital radius is measured

Velocity vector
of a full Moon

Velocity vector
of a new Moon

Velocity vector
of Earth

The Sun

Fig. 5B.1 The configurations for the Earth-Moon-Sun system requiring the least amount of energy to blow the Moon clear of the Sun (full Moon), and the most (new Moon).

how much of a kinetic boost does the Moon need to make its total energy, relative to Earth, a positive value.

The energy yield of all the exploding nuclear waste has to exceed $+3.77 \times 10^{28}$ joules. That equates to the equivalent of nine trillion (9×10^{12}) megatons of TNT, 692 trillion (6.92×10^{14}) times the energy released by the Hiroshima bomb, or 155 billion (1.55×10^{11}) times the explosive yield of the largest nuclear device ever exploded on planet Earth (the Soviet Union's Tsar Bomb at a whopping 58 MT). Using $E = mc^2$, it would require the equivalent of converting 416 billion kg of matter (equivalent to a small asteroid) directly to energy.

Our Moon moves through more than one gravity well, however, and the real story of *Space: 1999* comes not from creating enough explosive thrust so that our Moon escapes Earth's influence, but rather the *Sun's*, so that it's free to wander the Galaxy. Performing the calculation for leaving the Sun's influence is a little more difficult, since that number depends upon where the Moon is in its orbit relative to Earth. The maximum amount of thrust required would be when the Moon is between the Sun and Earth, what is known as a "new" Moon. In that case it's closer to the Sun, hence it feels the pull of gravity more. Its velocity is also in the opposite direction of Earth's, so relative to the Sun it's moving more slowly. The scenario looks like this (Fig. 5B.1):

So the values for the energy calculations would be:

$$r = r_{Earth} - r_{Moon}$$

$$v = v_{Earth} - v_{Moon}.$$

Since Earth's orbit is very nearly circular ($r_{Earth} = 1.49 \times 10^{11}$ m), we'll assume its velocity is constant also ($v_{Earth} = 29{,}662$ m/s). (Technically, velocity is a *vector*, that is, a physical quantity that has a magnitude pointing along some direction. As Earth moves around the Sun, gravity tugs at it, constantly changing the direction of its velocity to keep it flying off into space. The magnitude—how fast it is moving—remains about the same, however. For this example though, we only have to consider the magnitudes of the various velocity vectors because we are looking only at situations where the Moon's velocity vector is parallel to the Earth's, which allows us to use simple addition and subtraction.)

When Earth is between the Moon and the Sun, the Moon is farther from the Sun, and feels its gravity less. Also, it is now moving in the opposite direction as when it was a new Moon, so its velocity would be added to that of Earth's, giving us:

$$r = r_{\text{Earth}} + r_{\text{Moon}}$$

$$v = v_{\text{Earth}} + v_{\text{Moon}}.$$

In order to calculate how much energy it would take for the Moon to escape Sol, we apply the same energy calculations as we did for escaping Earth. We find that it takes, between 7.4 quintillion (7.4×10^{15}) and 8.3 quintillion (8.3×10^{15})—a factor between 830 and 960 times—more 1 megaton nuclear warheads than it did for Earth in order for the Moon to escape Sol and wander the Galaxy[29]. That's roughly three times the energy yield resulting if the Earth-threatening asteroid Toutatis was converted directly into energy. Or, if those 1 MT nuclear warheads were all the size of basketballs (unrealistically small for 1 MT devices), at minimum they would fill a sphere with a radius just smaller than four Earth radii (6371 km).

In summary, this one is far from plausible, so we'll chalk it up to the realm of story conceit and serious suspension of disbelief.

---

[29] Since the Moon's phase was a waxing crescent on September 13, 1999, the day it was blown free of the Solar System as per *Space: 1999*, the energy it took would have been closer to the 8.3 quintillion MT end of that range.

6
Radiation: An All-Time Glow

Radiation? What's a little radiation?
The Doctor, *Doctor Who*, "The Twin Dilemma"

I don't want to be human! I want to see gamma rays! I want to hear X-rays! And I want to… I want to smell dark matter!
Brother Cavil, *Battlestar Galactica*, "No Exit"

A day without electromagnetic radiation is like a day without sunshine.
Anonymous

In very early science fiction films, when mad scientists were commonplace, the Insane Genius Weapon of Choice was typically electricity. The public, ignorant of the scientific properties of electricity viewed it with shock[1] and awe, and more than a little bit of fear. They believed it had deadly and unpredictable properties that bordered on the mystical (something exploited by charlatans and well-meaning doctors alike, who promulgated a variety of electrical-based "cures" for practically every disease under the sun). By the 1950s and 1960s, the cinematic phenomenon that created Godzilla (or Gojira)[2], woke Gamera[3], created giant spiders[4], ants[5], and grasshoppers[6], and even gave a brachiosaurus the power to deliver electric shocks[7,8] was another phenomenon that the public had come to believe had deadly and unpredictable properties bordering on the mystical: *radiation*.

---

[1] See what we did there?
[2] Or, more recently, was used in a futile attempt to destroy Godzilla.
[3] *Gamera* (1965) A nuclear blast thawed Gamera, who was frozen in ice. In the 1995 reboot, Gamera was a genetic engineering product of Atlantis.
[4] *Tarantula* (1955).
[5] *Them!* (1954).
[6] *Beginning of the End* (1957).
[7] *Behemoth, the Sea Monster* aka *The Giant Behemoth* (1959).
[8] Nice how they covered all their angst-inducing bases with radiation *and* electricity.

Radiation lurks like a scientific Boogey Man[9] on the edge of our imaginations, and indeed, no end of scriptwriters have used it as simply a scare device. Say "radiation," and the hearts of many skip a beat. Why shouldn't it? The public has been conditioned to feel that way by the pejorative use of the term in the science fiction B-movies of the 1950s, and by the media storms that surround very serious real-world events like Hiroshima, Chernobyl, the Windscale fire, Three Mile Island, and Fukushima. Still, the same people who suffer an adrenaline rush at the mere mention of radiation will happily discuss its ills over a frozen dinner that was heated by microwave radiation. Radiation is essential to our existence. Life on Earth would not have evolved were it not for radiation. There is, perhaps, no phenomenon in science that has been more misunderstood, or whose effects have been more inaccurately portrayed—or at least wildly exaggerated—than radiation.

Part of the problem is that the term *radiation* is actually a very nonspecific word—it is an umbrella term that does not say much in itself, just like the phrase "life form" doesn't tell a crew beaming down to an alien planet whether they are about to encounter harmless microscopic amoebas, or a giant, multitentacled, predator. It derives from a Latin term meaning "glittering" or "shining," but there are two fundamental types of radiation—one composed of subatomic particles and the other of electromagnetic waves—which have very different properties. Within those, physicists, have named many subtypes of radiation: Hawking radiation (produced at the event horizon of a black hole), Cherenkov radiation (which is a sort of optical shockwave, similar to a sonic boom, that gives high-energy nuclear waste stored underwater an eerie blue glow), Bremsstrahlung radiation (literally, "braking radiation," which is produced when an energetic charged particle like an electron is slowed down or deflected), neutron radiation (a big problem in current fusion reactor designs)... the list goes on and on. The only commonality between these types of radiation is that something—whether it is subatomic particles or electromagnetic waves—moves radially away from its source. Grouping these phenomena together under one blanket term is equivalent to lumping planets and stars into a single term that means "glowy things in the night sky." While this term would be an accurate description of stars and planets, it omits many context-changing details. A basic understanding of radiation helps us understand the physics behind some of the technology and weaponry, along with many of the astrophysical phenomena, depicted in fictional science productions.

---

[9] Since the topic of scary insects has already been raised, the root of "Boogey" is the Welsh term *bwg* (pronounced "boog"), which is the root of the modern word "bug". So when Mr. Oogie Boogie in *The Nightmare Before Christmas* (1993) is depicted as a burlap sack full of insects, that is actually an awesome allusion to the fact that the Boogey Man is really the "Buggy Man".

6.1 Particulate Radiation

Particulate radiation typically occurs when subatomic particles are released or ejected from unstable atomic nuclei, or from fission or fusion processes involving the interaction of multiple nuclei. Often particles are emitted with very high energies.

Previously we explored how the number of protons within an atom's nucleus dictates that element's chemical properties by controlling the number of electrons available to interact with other atoms, but the count of neutrons in a nucleus can vary considerably among isotopes of the same element. Some nuclear combinations of protons and neutrons are so unstable as to be practically impossible—they decay into smaller pieces instantaneously. Some isotopes are *metastable* and exist for a period of time—ranging from fractions of a second to millions of years—before spontaneously decaying into daughter products. Metastable isotopes that emit radiation through the process of decay, are said to be *radioactive*, and are known as *radioisotopes*.

Tritium, or hydrogen-3 (which is hydrogen with two extra neutrons) and carbon-14 are well-known radioisotopes, as is aluminum-26 (which plays a key role in the early evolution of planets, as well as the 2007 *Eureka* episode "Sight Unseen"). Patients with suspected thyroid conditions drink iodine-123 to assist with medical imaging, and there is an entire suite of radioisotopes, called radiopharmaceuticals, used in medicine[10]. On the darker end of the lethality scale, polonium-210 made the news in 2006 as the radioisotope that killed former Russian spy Alexander Litvinenko. All isotopes of uranium and plutonium are radioactive.

When physicist Ernest Rutherford and Paul Villard discovered the three most common forms of radiation, they labeled them "alpha rays," "beta rays," and "gamma rays"[11], based upon their penetrating power. Alpha particles—two protons bound to two neutrons (which are in effect the naked nuclei of helium atoms, stripped of their orbiting electrons)—can be stopped by clothing, a sheet of paper, often even the dead outer layer of the skin. Alpha particles can not even travel more than a few feet through air. The americium-241 in smoke detectors is an alpha emitter, and the detector works because even a whiff of smoke is enough to block a notable number of alpha particles. Alpha particles are only dangerous if you inhale or swallow a substance that emits them, such as the radon gas that can accumulate in basements.

Beta particles are merely electrons that are unbound to any atom. Beta emitters come in two varieties. Some radioisotopes that produce beta radia-

---

[10] Radioactivity can be life-saving, in the proper context.
[11] The very same gamma rays that irradiated Bruce Banner to create the Hulk. We're getting there.

Fig. 6.1 On the *left* is the traditional "trefoil" symbol used to identify ionizing radiation hazards. Because the degree of actual hazard involved covers a wide range from minimal to highly dangerous, in 2007 the International Atomic Energy Agency introduced the symbol on the *right*. It is specifically designed to be put on vessels or devices containing radioactive sources that produce lethal amounts of radiation, and is placed on housings so as to be visible only if someone is attempting to expose the source. If you ever see the symbol on the *right* and you're not a nuclear engineer, follow the example of the stick figure and run—not walk—away. If you *are* a nuclear engineer, double check to make sure you know what the hell you're doing.

tion, like tritium and carbon-14, emit low-energy particles that are stopped easily. Even high-energy beta particles, the kind that are generated by potassium-32 and strontium-90, can be stopped by a thin sheet of metal foil (but they can penetrate the skin, so it's not a good idea to handle a beta emitter directly).

Gamma rays are a form of electromagnetic radiation, and can penetrate very deeply, or even pass through, most matter. They can do a lot of damage to biological tissue, which is a reason no-one but the Hulk's mild mannered alter ego Dr. Bruce Banner was willing to self-experiment with gamma rays. They are stopped by lead shielding, and we'll come back to gamma rays in more detail later (Fig. 6.1).

Some radioisotopes emit neutrons when they decay. Free neutrons are also a byproduct of both fission and fusion reactions. Depending upon how much kinetic energy they have, neutrons are called *thermal* (slow) *neutrons* or *fast neutrons*. Neutron radiation is used to initiate and maintain fission chain reactions in nuclear power plants and nuclear weapons. Some isotopes of uranium and plutonium, when struck by neutron radiation will decay into lighter nuclei, releasing energy in the form of gamma rays as well as more neutrons— neutrons that can interact with other nuclei to cause them to decay as well. Moderate this *chain reaction* so that it proceeds at a slow rate, and you have a nuclear reactor; allow the process to propagate unimpeded, and you have either a meltdown or a nuclear weapon. It should be noted at this point that even meltdowns like Chernobyl and Fukushima that blew the roof off their buildings didn't produce *nuclear* explosions, although they did release a significant amount of radiation. It turns out that one of the hard parts of making a nuclear weapon go *boom* properly is getting the chain reaction to propagate through the bomb faster than the energy that is being released can blow it apart. If the core of the bomb comes apart before the chain reaction can really get started, you have what weapons scientists refer to as a *fizzle*, which is what many people believed happened when North Korea tried to conduct its first nuclear weapons test in 2006.

Like gamma radiation, neutron radiation is very penetrating, but unlike gamma rays, which are stopped only by significant amounts of matter like lead shielding, neutron radiation can be blocked by lighter elements like water, which explains why nuclear reactors are surrounded by water—as a coolant and as a neutron *moderator* (a moderator is any substance used to control neutrons so that they stay in an optimum kinetic energy range for nuclear reactions).

While there is a veritable zoo of subatomic particles, there is one form of particulate radiation that can be emitted from fusion in the core of a star, radioactive decay, or high-speed particle collisions, which warrants special mention: *neutrinos*. The word "neutrino" is Italian for "little neutral one." Not only do neutrinos carry no charge, they have a nearly-undetectable amount of mass, and travel at speeds very close to that of light.

Because neutrinos have little mass, travel at such high speeds, and have no charge, they rarely interact with normal matter. Not only does this make them difficult to detect (because detectors are composed of normal matter), they also zoom through large amounts of matter, like planets, unperturbed.

It is because of this ability to travel through solid rock in the blink of an eye that stock brokerages have been investigating whether it is possible to build a "neutrino gun" that can send a beam of modulated neutrinos through Earth to a receiver on the other side. If it was possible that such an information-carrying signal could be sent through Earth, this would give brokerages a 30–40 millisecond advantage over more standard forms of communications, like satellites or fiber optic cables. In the highly competitive world of high frequency stock trading, 30 milliseconds is a huge advantage, so in the long run it could pay for the (estimated) $1 billion development cost[12]. The chief difficulty is, of course, building a detector that can sense the neutrinos efficiently.

In the 2007 episode of the reimagined *Battlestar Galactica* series entitled "Rapture," *Galactica* orbits an Earth-like planet, dubbed the Algae Planet, which is orbiting a star nearing the end of its life. The writers of the episode

---

[12] http://www.forbes.com/sites/brucedorminey/2012/04/30/neutrinos-to-give-high-frequency-traders-the-millisecond-edge/.

needed a precursor that would give the characters warning—but not *too* much warning—that the star they are orbiting is about to explode.

Lt. Gaeta reports that *Galactica*'s sensors have detected a *helium flash* within the star, signifying that the star is in the early stages of going supernova. In fact, a helium flash occurs only when stars similar to our Sun reach the end of their first red giant phase[13], and are transitioning from fusing hydrogen as a primary power source to fusing helium. A helium flash actually signifies a transition to a plateau of stability, not an impending cataclysm.

Ideally, from a physics standpoint, Lt. Gaeta would have been looking for a surge in neutrinos, which are emitted in titanic amounts in the early stage of a supernova explosion. Neutrinos would have fit the writers' wishes perfectly, except for two problems. First, there would be little reason for a warship like *Galactica* to have a neutrino detector. The writers could probably have come up with a reason why *Galactica* has a neutrino detector. The second reason is the real biggie: Every second, over 50 trillion neutrinos generated by fusion in the core of our star pass through you. At this number, somewhere from once per week to once per lifetime (precise estimates are difficult), a neutrino undergoes some form of interaction with an atom in your body. Even affecting one atom a day would be a rate that is way too low to be biologically significant. Yet, the they are emitted in staggering amounts during a supernova, despite interacting weakly, and very infrequently, with normal matter, the sheer number of neutrinos would have killed everybody aboard *Galactica*, as well as those on the planet below. The human race as we know it would have perished at the Algae Planet[14,15].

6.2 Electromagnetic Radiation

If mere mention of the word "radiation" gives you a case of the screaming heebie-jeebies, you might want to sit down. You're emitting radiation right now, and so is everything around you. You're bathed in radiation, and you've been taking that bath for your entire life. Although you are reflecting sunlight or artificially-created light in the room in which you are sitting, you are also

---

[13] Our star will become a red giant twice before ending its life.

[14] The mean free path—the average distance between successive collisions—of a neutrino passing through lead is approximately 1.5 light years. That value is for low-energy neutrinos, the kind released by radioactive decay. Higher energy neutrinos from supernova explosions will "only" pass through 7 light weeks (about 285 times the distance between the Sun and Neptune) of lead. So neither *Galactica* nor the Algae Planet would have shielded our heroes.

[15] In the *Star Trek: The Next Generation* episode "The Enemy", viewers learn that Geordi LaForge's VISOR (the device that allows him to see) can detect neutrinos. Not likely, even in the twenty-forth Century.

generating electromagnetic radiation simply by virtue of the fact that you are at a higher temperature than absolute zero.

Electromagnetic, or EM, radiation is energy also emitted by an atom or molecule when it undergoes a transition from a high-energy or excited state to a lower-energy state. We have described how the electrons in the Bohr-Rutherford Model orbit atomic nuclei similarly to planetary systems, but there is one key difference we have yet to explore. In a planetary system, a planet could, in principle, orbit *any* distance from its central star. Electrons orbit at specific distances[16], with each orbit corresponding to a particular amount of energy that an electron can possess. (Which is to say that an electron's energy in any given atom must be one of a finite number of discrete energy levels.)

Now, just because an electron is sitting in a specific orbit does not mean that it is constrained to remain in that orbit forever. Collisions between atoms or with other subatomic particles, the presence of an electrical current, even the absorption of light, can give an electron a kick of kinetic energy; forcing it into a more distant, higher energy, orbit. Give an electron enough energy, and you strip the electron from the atom entirely, a process known as *ionization*.

Electrons, however, prefer to reside in the lowest energy state possible, so if left alone they spontaneously drop from excited states to lower-energy orbits. The electron may jump straight back to the orbit from whence it came, or, like the rungs of a ladder, may return piecewise—falling from one orbit, to the next, to the next, until it is in its lowest energy state. Recall, though, that energy can neither be created nor can it be destroyed. If the electron was boosted to a higher energy orbit, then drops to a lower energy orbit, where did that energy go? It zoomed away from the atom, literally at the speed of light, as a tiny packet of energy called a photon—the fundamental component of electromagnetic radiation[17].

Photons are, essentially, tiny packets of potential energy. Every form of EM radiation—from radio waves to visible light to gamma rays—is nothing but photons of different energies. So when we say that the theoretical fastest attainable speed, 3×10^8 meters per second, is the "speed of light," that term, more correctly, is the "speed of electromagnetic radiation in a vacuum." "Speed of light" is much easier to say, to the point that even *The Big Bang Theory*'s perfectionist Sheldon Cooper says, "speed of light." We'll go with that.

---

[16] We're stretching the word "distance" to breaking point here, since the shapes of electrons orbits can be very complex, with all kinds of three-dimensional lobes and things, rather than the simple two-dimensional ellipses of a planetary orbit, but it'll do for our purposes.

[17] While there are other ways in nature to absorb or emit photons, this is the most common. In water, the two hydrogen atoms act as if they are attached to the more massive oxygen by springs. It is possible for the molecule to absorb electromagnetic energy to increase the rate of this vibration, or the kinetic energy of this *vibrational mode*. Exciting this vibrational mode in water is, in fact, how a microwave oven heats your food.

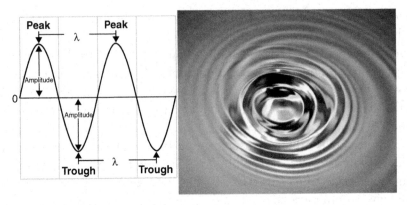

Fig. 6.2 The components of a wave, and an example—surface waves in water. (Photo by Roger McLassus)

While light can be considered a stream of particle-like photons, it also has wave-like properties, and behaves like a wave of varying electric and magnetic fields. We see the components of a wave in (Fig. 6.2):

The point of maximum height of a wave is called its *peak*, or crest; and the point of the maximum depression, a *trough*. The maximum height or maximum depression of a wave it is called its *amplitude*. The distance from one crest to the next crest, or one trough to the neighboring trough, is called the *wavelength*, typically denoted by the Greek letter λ (pronounced "lambda."). The wavelength is a physical length, and is measured in meters (ham radio operators will talk about working the "20-meter band" on their radios, for example), but for an electromagnetic wave, its not unusual to see wavelengths given in nanometers (billionths of a meter) or Ångstroms (10 billionths of a meter) for things like visible light or X-rays, respectively.

The number of peaks or troughs that pass a measuring instrument or observer in a given amount of time is called the *frequency*, and the unit of frequency is cycles per second, or *Hertz* (abbreviated Hz). Because light travels at a constant speed, frequency and wavelength are inversely related for EM waves. A photon with a small wavelength has a high frequency, and a photon with a long wavelength has a low frequency. Think of water waves passing a stick. If the wavelength is a meter long, it will take several seconds for the water level at the stick to rise, fall, and rise again. If the wavelength is just a couple of centimeters, the water level could cycle through rising, falling, and rising again several times in just one second.

Sometimes frequency is measured in millions of cycles per second: megahertz, or MHz. The FM dial on your automobile radio (if you still have a radio) is also in MHz, while the AM dial is in units of thousands of cycles per second, or kilohertz (kHz). Communicating using radio waves requires picking a particular frequency, with the sender and receiver tuned to the same

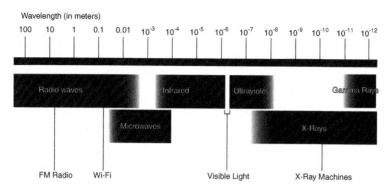

Fig. 6.3 The electromagnetic spectrum, denoting the different spectral ranges, from radio to gamma rays

frequency[18]. In *Contact* (1997), Jodie Foster's character, Dr. Ellie Arroway, is able to discover a signal from an alien civilization by guessing[19] the frequency they are using—4.4623 GHz[20]—and tuning a very powerful radio telescope to that frequency.

The energy of a photon is directly related to its frequency. The higher the frequency (or lower the wavelength), the higher the energy. Electromagnetic energy comes in many "flavors", corresponding to photons of characteristic energy ranges. The entire gamut of EM radiation—from the low frequency of radio waves to high frequency gamma rays—is called the electromagnetic spectrum, as shown in Fig. 6.3 (note that the diagram uses a logarithmic scale, which means each division is actually one tenth the size of the preceding division).

Radiation at the low frequency/long wavelength end of the EM spectrum—with wavelengths greater than a few centimeters (frequencies less than about 300 MHz)[21]—are typically called radio waves. Today's technology relies heavily on the radio portion of the EM spectrum. These are the waves that broadcast antennae modulate to transmit music, sports, and news to your AM (540–1710 kHz)/FM (87.5–108.0 MHz) radio receiver, and broadcast television to your rabbit ears[22,23] (30–300 MHz for channels 2 through 13;

---

[18] Very early radios used a huge range of frequencies simultaneously, but this was enormously inefficient for long-distance communications and very prone to interference from natural radio sources and other human-made transmitters. Nonetheless, for modern high-speed, short-range communications there is renewed interest in so-called *ultrawideband* communications, in which the signal is transmitted across a large band of frequencies and stitched together digitally.

[19] It was not a random guess, it was a very educated guess, as we'll explain in the next volume of *Hollyweird* Science.

[20] GHz means "gigahertz," or billions of cycles per second.

[21] The exact boundaries for the different segments of the EM spectrum vary slightly among scientific disciplines.

[22] You just said either, "Huh?" or "Boy, I feel old."

[23] With the transition to digital television broadcasting, using an antenna to watch TV is actually having something of a renaissance—the interference problems that plagued analog TV reception is much

300–1000 MHz for channels 14 through 69). The carrier signal for Sirius satellite radio is at 4 MHz.

Another important application of EM radiation is RADAR. The term RADAR is an abbreviation that stands for RADio Detection and Ranging. To establish the distance to objects like ships or aircraft, the simplest RADAR system uses an antenna to broadcast a pulse of EM waves in the radio range of the spectrum. Radio waves are reflected back by a ship or by an aircraft—they are reflected particularly well by metals—and received by the same antenna that sent the pulse. The system measures the time it takes for that pulse to be reflected back and, knowing the speed of light, it is a simple matter to calculate the distance (more sophisticated systems use a continuous RADAR beam instead of individual pulses to reduce false signals, but the details of how they work are beyond the scope of this book).

Sound waves are unrelated to EM waves—they are of a fundamentally different nature and require a medium through which to propagate, so they do not travel through space, unlike EM waves. Because sound is a wave, an acoustic wave, it does share some properties with EM waves, notably concepts like frequency and wavelength. Humans are sensitive to acoustic waves ranging from about 20 Hz to 20 kHz: for radio and television broadcasts, the signal that carries the broadcast, known literally as the *carrier signal*, is of a much higher frequency than the audio range. The electronics of the radio receiver step that down to audible frequencies and convert the electromagnetic signal into an acoustic wave (and a video stream if present).

In shows from the early days of televisions, or period productions, you may see a character tune an AM radio and, amidst the static between stations, there are sharp whistling sounds[24]. These are literally called *whistlers*. Lightning generates electromagnetic waves in the 1–30 kHz range, a range that, if converted to acoustic waves of the same frequency, overlaps with the range of human hearing: the highest amplitude emissions are in the 3–5 kHz range—on a piano that would span F7# to D8# (if there was such a thing as D8# on a piano keyboard).

Planets can also emit radio waves. When electrons are accelerated, or their trajectories altered, by strong magnetic fields, they emit a form of radiation called *synchrotron radiation*. Electrons trapped in the strong magnetic field of gas planets like Jupiter emit a form of synchrotron radiation, called *decimetric radiation* in the 100 MHz–15 GHz range of the radio spectrum.

---

reduced with a digital signal, and the picture quality can be better than that transmitted over cable or satellite, as cable and satellite companies often compress the video signal.

[24] You may even recall this yourself. One of the authors does. He's older than he looks.

This phenomenon was used in the reimagined *Battlestar Galactica* 2007 episode "Maelstrom," as the RagTag Fleet used the synchrotron radiation of a gas giant planet to mask their presence from the Cylons while *Galactica* refueled and resupplied other ships of the fleet. When Starbuck believes she has spotted a Cylon craft, Colonel Tigh remarks that it will be difficult to get a positive ID with all the synchrotron radiation present. All this suggests that the DRADIS (the equivalent of RADAR in the *Galactica* universe) for both *Galactica* and the Cylons operates within the radio portion of the spectrum.

Photons slightly higher in energy, in the range of approximately 1 meter (300 MHz)–1 mm (300 GHz–0.3 THz[25]), are known as microwaves—by some definitions of the boundaries of the EM spectrum, microwaves are considered part of the radio portion of the spectrum. These are the very microwaves that reheat last night's pizza, or boil the water for your morning coffee, in a microwave oven. Water molecules, as well as certain plastics, preferentially absorb microwaves with a wavelength of 12.2 cm (2.45 GHz). So although

---

The radiation left over from the Big Bang is the same as that in your microwave oven but very much less powerful. It would heat your pizza only to −271.3 °C—not much good for defrosting the pizza, let alone cooking it.

Stephen Hawking, Theoretical Physicist

---

photons at this wavelength are not highly energetic individually, because they are so readily absorbed by water molecules, they efficiently heat food.

It turns out that the concepts of "transparent" (an object stops no light), "translucent" (an object stops some light), and opaque (an object stops all light) are highly dependent upon the frequency of the radiation. While typical glass is transparent to EM radiation of visible wavelengths, it is translucent, or even opaque, to ultraviolet radiation (which we will visit in a bit) of different frequencies. Space shuttle hatches were fitted with a single small window composed of quartz—transparent to ultraviolet radiation—to allow UV observations. When not in use, the window was covered to avoid giving passing crew sunburns.

In a similar vein, most fabrics and some plastics are transparent to millimeter wave radiation (which actually ranges between 1 and 10 mm), so it is the portion of the EM spectrum to which today's airport full body scanners are sensitive. Many materials that could be used in weapons, not simply metals, have unique signatures in the millimeter band as well, implying that scanners can not only see through clothing, they can potentially pinpoint weapons that passengers are carrying.

---

[25] The value of 1 Terahertz (THz) is 1 trillion Hz.

There are two types of millimeter wave scanners: active and passive. Passive scanners compute images using just the radiation naturally emitted by a human body or other objects being scanned. Active scanners act more like a RADAR, emitting millimeter wave radiation at the subject, and constructing an image by measuring the reflected signal.

The Transportation Security Administration (TSA) web site says:

> TSA currently uses millimeter wave AIT[26] to safely screen passengers for metallic and nonmetallic threats, including weapons and explosives, which may be concealed under clothing without physical contact to help TSA keep the traveling public safe.

> All millimeter wave AIT units deployed at airports are outfitted with software designed to enhance passenger privacy by eliminating passenger-specific images and instead auto-detecting potential threats and highlighting their location on a generic outline of a passenger that is identical for all passengers.

If this sounds strangely familiar, the full-body scanners of today were presaged by the 1990 film *Total Recall*, where the film depicted a full-body imaging system that not only saw through clothing, but also identified Douglas Quaid's weapon (Fig. 6.4).

Microwaves at other wavelengths are routinely used around us in the open air. The RADAR guns employed by law enforcement to measure speed of vehicles on our roads broadcast in the microwave portion of the spectrum. Look closely at the label on a WiFi base station, and often the operational frequency is listed as 2.4 GHz, or 2.4 gigahertz. This is also in the microwave portion of the spectrum. Microwaves are also used to carry spacecraft communications.

Ranging from 1 mm[27] (300 GHz) to 720 nm (416 THz), even higher in energy still, is infrared (IR) radiation. TV remote controls generally operate in the infrared, either at 870 nm or in the range from 930 to 950 nm. Terahertz radiation, often called the submillimeter band by the military and by astronomers, is at the low-frequency end of the infrared portion of the spectrum, on the border with microwaves: from 1 μm (0.3 THz) to 100 μm (3 THz).

Infrared radiation is outside of the range of wavelengths the human eye can see, but not beyond the range a cell phone camera can see. Put a cell phone in camera mode, and shoot a remote control into the camera lens. The cell phone screen will show a purple glow from the transmitter of the remote control.

Infrared radiation is sometimes called thermal radiation. Hot objects radiate a great deal of energy in the IR portion of the spectrum, and there is a

---

[26] Advanced Imaging Technology.
[27] One nm is a nanometer, or 1 billionth of a meter.

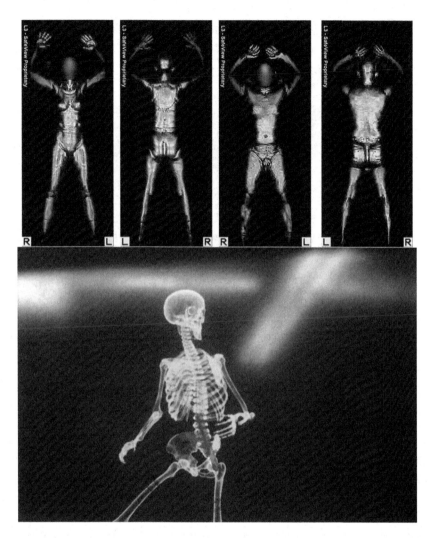

Fig. 6.4 Security scanners that can see through clothing and automatically identify potential threats were presaged by the film *Total Recall* (1990). The *upper panel* shows the type of image possible with real millimeter wave scanners. In the *bottom panel*, Quaid's concealed weapon is detected and flagged by a security scanner. (Still Copyright © Tristar Pictures. Image credit: Transportation Security Administration and PhotoFest)

subrange within the infrared called the *thermal infrared*, or long-wavelength infrared, radiation from 15,000 nm (15 microns or 20 THz) to 8000 nm (8 microns or 37.5 THz).

In the 1996 movie *Eraser*, starring Arnold Schwarzenegger and Vanessa Williams, the co-star was a DoD-contractor-developed rail gun that could be fired by individuals like a rifle. The Hypervelocity Mag Pulse Prototype also had a targeting system that allowed the user to see through walls to acquire a

Fig. 6.5 FLIR (forward-looking infrared) image of Boston Marathon terrorist Dzhokhar Tsarnaev hiding in a covered boat. (Photo credit: Massachusetts State Police)

target. Counterintuitively, this imaging system is far more scientifically plausible than a personal rail gun that fires "aluminum rounds at almost the speed of light."[28]

Just as millimeter wave radiation can pass largely unimpeded through most clothing, the phenomenon that is the basis for airport scanners, some radiation in the thermal IR part of the spectrum passes easily through many solid objects (like fabrics, vegetation, or, in some cases, even walls). Many military and law enforcement aircraft are mounted with forward-looking infrared, or FLIR, cameras that see in this range of the spectrum. The world saw an example of FLIR when, on April 20th, 2013 the Massachusetts State Police released a FLIR image acquired by helicopter of Boston Marathon bomber Dzhokhar Tsarnaev hiding in a covered speedboat sitting in a driveway in Watertown, MA (Fig. 6.5).

Objects significantly hotter than people emit radiation in the mid-IR range, from 8000 to 3000 nm (3 microns or 100 THz). Heat-seeking, or IR-seeking, missiles operate in this range—they are programmed to home in on the IR radiation emitted from the hot tailpipes of an adversary's aircraft. Some more sensitive heat-seekers can even home in on the heat generated by the friction of the aircraft's frame with the air molecules through which it flies.

---

[28] There are some serious conservation of momentum and relativity issues here.

The term "light" is often used as a blanket term for all forms of electromagnetic energy, but in the strict definition of the term, it means electromagnetic energy visible to the human eye. That corresponds to a very narrow portion of the EM spectrum from red at the low-frequency end spectrum 720 nm (416 THz) to violet (770 THz/390 nm) at the high-frequency end.

Since the frequency of light is directly proportional to the energy of the photons that transmit it, "color" simply means that the receptors in human eyes, the cones, respond differently to photons of different energies. Considering colors quantitatively, as a measure of energies, is not inherently how one typically thinks of color, and this puts the concept of "color" into a different … light.

Keeping in mind that the frequency of a color corresponds to a photon's energy, it is worth discussing two physical laws that describe the nature of radiation emitted by objects as they are heated: *Stefan's Law* and *Wien's Law*. These laws are simply formal quantitative descriptions of phenomena that everybody has observed—either first-hand or on screen.

Any object whose temperature is above absolute zero radiates thermal energy or *thermal radiation*. Since temperature is a measure of the average kinetic energy of a substance—a measure of how fast the atoms are moving—the hotter an object is, the faster its atoms move. As atoms collide with one another, charged particles that make up atoms—mostly electrons—are constantly being accelerated and decelerated as the atoms are alternately jolted and braked[29]. If you wiggle a charged particle in this way it will produce electromagnetic radiation (a similar effect is used in the antennae attached to radio transmitters to produce radio waves). EM radiation can also be produced when an electron is excited to a higher level than it normally occupies, and then falls back to its lower level—in the process it emits a photon, a phenomenon that is a key part of how lasers generate their light beams. If atoms are moving faster, they change direction more frequently, and collisions between atoms are increasingly likely—the kind of collisions that kick electrons into higher energy orbits, or excited states. Radiation emitted in this fashion is called thermal radiation, and the warmer an object, the more radiation it produces.

This emitted light is different than the color an object reflects. Take green plants, for example. Plants are green because sunlight, thermal radiation from the Sun, impinges upon them, and green is what is reflected to your photoreceptors—your eyes. Plants are also emitting thermal radiation in the infrared. So are you. Your warm body is emitting infrared waves right now. In others words, you are emitting radiation all the time!

---

[29] The emitted radiation follows a distinct pattern called the *blackbody curve*. Figuring out the reason for the shape of the blackbody curve was a major step in the development of quantum mechanics.

Stefan's Law simply states[30] that the hotter an object is, the more photons it emits at all wavelengths, therefore the brighter it is. This is easy to see using an incandescent light attached to a light dimmer[31]. Incandescent lights work because electrons passing through a tungsten filament heat the filament through, essentially, friction. Dim the light as much as possible, very little current flows through the filament in the bulb, very little friction, very little light. Turn up the light, allow more current to pass through the bulb, more current flows, the filament heats, and it glows brighter[32].

Wien's Law states, essentially, that the peak wavelength where an object emits most of its thermal radiation is determined by its temperature, and as the temperature increases that peak shifts to higher and higher energy (or lower wavelength). Tympanic ear thermometers are based upon this principle; they determine body temperature by measuring the thermal radiation within the ear cavity, finding the wavelength of maximum intensity, and applying Wien's Law.

Combining Stefan's Law and Wien's Law, imagine an iron put into a blacksmith's forge. At room temperature, it is emitting mostly infrared radiation, but you can not see this. As the iron heats past 525 degrees Celsius it will start to glow red. As it gets hotter, not only will it glow ever brighter, but it will turn orange, then yellow, eventually bluish-white. Heat it to extreme temperatures, and it will be "white hot."

This is why infrared cameras can be used to find people in the dark or at night. People literally glow in infrared. The ambient background glows in the infrared as well, but a warmer object—like a person—emits more radiation, i.e., it is brighter, at all wavelengths. Some night vision scopes work on this principle (others simply amplify the tiny amount of available light in the visible portion of the spectrum). This also explains why the alien antagonist in the 1987 *Predator* movie was such a lethal hunter. It had infrared vision, and was able to track its prey through the jungles: only when Arnold Schwarzenegger's character, Dutch, masked his infrared signature by hiding in relatively cool mud was he able to counterattack effectively. Of course, if the temperature of the jungle was the temperature of a human body or higher (which is reasonable for a jungle), Dutch and his team would have been naturally camouflaged already.

---

[30] We will spare you the math for the moment.

[31] This is more directly applicable to an incandescent bulb, not a fluorescent bulb. There is a parallel with fluorescent bulbs, but that is a bit more involved.

[32] The overwhelming majority of the radiation of incandescent bulbs is emitted in the infrared, in a region of the spectrum where people can not see. It is wasted energy. Fluorescent bulbs are "tuned" so that the majority of their emission is in the visible portion of the spectrum, meaning they emit more visible light while using far less energy. This explains why; when you go to the hardware store, fluorescent bulbs have two output wattages listed: their actual output, and their incandescent equivalent.

Science Box: Hot and Cool Lighting in Cinematography

One of the most direct impacts of Wien's Law on Hollywood is not found within dramatic narrative, but rather in bringing it to the screen, in the lighting and cinematography. Many a budding young cinematography student has been confused that a high color temperature of over 5000 K (outdoor lighting is typically 5600 K, close to the temperature of the surface of the Sun) is considered "cool" because it has a blue cast, while color temperatures around 3000 K (indoor lighting) are considered "warm" because that light, light typically emitted by a hot tungsten filament (3200 K), the sort one finds in incandescent light bulbs, is reddish. How can a lower temperature correspond to "warm," and a higher temperature mean "cool"? In response, many a cinematography instructor has offered, "It doesn't make sense, that's just how it's done." In fact, it does make sense in the context of thermal radiation.

It only doesn't make sense initially because cinematographers, like the rest of us, are conditioned to view hot and cold by the "fire and ice" model—the same model by which cold water faucets are labeled in blue, and hot faucets in red. In this model, fire is red and hot, while ice is cool and blue[33]. Viewed through the lens of a thermal radiation model, the practice of blue = hot and red = cool makes perfect sense.

Figure 6A.1 shows the thermal radiation, or blackbody, curves for three objects, one at a temperature of 3000 K, one at 4000 K, and one at 5000 K. The figure shows

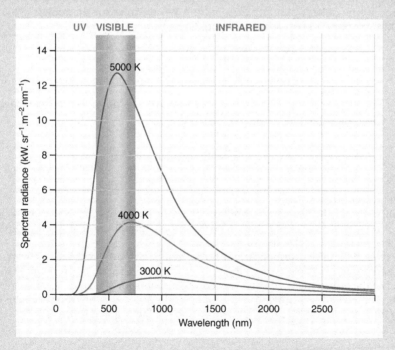

Fig. 6A.1 Thermal radiation curves for objects at 3000, 4000, and 5000 K. For hotter objects, the peak radiation output is at higher energies (*higher frequencies*). Hotter objects also emit more radiation at all wavelengths. (Image credit: Wikimedia Commons, user Darth Kule. Modified by user Cmglee and author KRG)

---

[33] Really cold water ice, the kind one can find in glaciers for example, actually is blue.

that, for hotter objects, the peak radiation output is at higher energies (higher frequencies or shorter wavelengths).

The shape of the thermal energy curves in Fig. 5A.1 are given by *Planck's law*:

$$B_\lambda(T) = \frac{2hc^2}{\lambda^5} \frac{1}{e^{\frac{hc}{\lambda kT}} - 1}.$$

Yikes! In this equation B stands for the energy output at each wavelength for a body at a given absolute temperature, λ is the wavelength, k is the Boltzman constant[34], h is Planck's constant[35], and c is the speed of light. That equation is overly complicated for our purposes, but it does lead to one that is useful: Wien's law or, more correctly, Wien's displacement law gives the wavelength (λ_{max}) where a hot object is radiating most of its energy, and is given by

$$\lambda_{max} = \frac{hc}{5} \frac{1}{kT} = \frac{2.9 \times 10^6}{T(\mathrm{K})} \mathrm{nm}.$$

Meaning that the wavelength in nm of the peak energy output (λmax) is a constant divided by the object's absolute temperature T. If we solve for the coolest (3000 K) and hottest (5000 K) temperatures in Fig. 6A.1, we get

$$\lambda_{max} = \frac{2.9 \times 10^6}{3000 \text{ K}} \mathrm{nm} = 967 \text{ nm},$$

which corresponds to the peak for the red curve, and is in the IR portion of the spectrum. Solving for the blue curve, the peak is at

$$\lambda_{max} = \frac{2.9 \times 10^6}{5000 \text{ K}} \mathrm{nm} = 580 \text{ nm},$$

which is yellow.

Tympanic (ear) thermometers work on this principle[36]. They scan the thermal energy present in the ear cavity, determine the wavelength of the peak energy output, and, knowing λmax, solve the above equation for T.

So blue definitely corresponds to a hotter temperature than red, and the terms are used properly in cinematography, at least from the standpoint of thermal radiation, not the entrenched "fire and ice" way of viewing hot and cold. Don't count on the hot and cold faucets being swapped any time soon, though; the only winners from that exercise would be the attorneys from all the scalding lawsuits.

---

[34] The Boltzman constant, 1.38×10^{-23} J/K relates the energy of a subatomic particle with its temperature.
[35] Planck's constant relates the frequency of a photon to its energy, and is 6.626×10^{-34} J s. We will discuss Planck's constant further in Chapter Seven: "A Quantum of Weirdness."
[36] They are also based upon spacecraft sensor technology.

Moving to even shorter wavelengths, and hence, higher frequencies beyond the visible portion of the spectrum, there is ultraviolet or UV—also known as black light, which has wavelengths that range from 390 nanometers (770 THz) down to 10 nm (30,000 THz or 30 PHz[37]). More energetic than violet "Ultra," literally translated, means "extreme," so this portion of the spectrum is "extreme violet"[38].

Ultraviolet light is often used for its ability to make some compounds *fluoresce*, that is, the invisible ultraviolet radiation stimulates the material to emit light at lower, often visible, frequencies. This is the operating principle behind fluorescent lights, where a mercury vapor is excited by the passage of electricity to produce UV light. A phosphor coating the light bulb fluoresces to produce visible light. Many bodily fluids, such as semen or blood, can be detected under a UV bulb thanks to fluorescence (as often portrayed by the forensic investigators on the crime procedural of your choice), although this can require a chemical[39] to be sprayed in the area to react with fluids before they will fluoresce.

UV radiation is also prized on-screen for its ability to kill vampires of the non-sparkly variety. In truth, UV light can really be a destructive, even lethal, form of electromagnetic radiation. It is used to disinfect water, since UV light will kill bacteria and render viruses inert by damaging their genetic material.

Fortunately our atmosphere—in particular the never-ending process of creating and destroying ozone, or O_3 molecules, in the stratosphere—absorbs most of the solar UV radiation incident upon Earth, radiation that would otherwise reach the surface. Still, even too much exposure to the small fraction of UV that makes its way to Earth's surface, can produce burns in the short term, and eye cataracts and skin cancers in the long term[40].

At even higher energies than UV are X-rays, from 10 nm (30 PHz) to 0.01 nm or 10 pm[41] (30,000 PHz or 30 EHz[42]). For laboratory applications, X-rays are generated by the decay of certain radioisotopes. X-Rays pass through soft tissues, but are absorbed by bone and other dense materials such as metal, leading to their ubiquitous use in medicine and dentistry to detect pathology in both skeletal and dental frameworks.

X-rays are also generated by highly energetic astrophysical phenomena such as planetary aurorae, pulsars, and black holes. *Accretion disks*, disks of mate-

---

[37] PHz stands for Petahetz, where 1 PHz is 10^{15}, or a quadrillion, Hz.

[38] In the 2008 *Eureka* episode "I Do Over" viewers were introduced to a type of EM radiation called "ultra indigo." Literally, this means "extreme indigo." The scientific term for that would be: violet.

[39] We've all probably watched enough crime shows that we've all heard of luminol by now.

[40] It's why the gingers and Irish among your authors have to apply SPF 1,000,000 before hitting the beach.

[41] 1 pm is 1 trillionth of a meter, or 10^{-12} meters.

[42] EHz stands for Exahetz, where 1 EHz is 10^{18}, or a billion billion, Hz.

Fig. 6.6 In the film *Interstellar* (2014) the supermassive black hole Gargantua was surrounded by an accretion disk of infalling material. An observer sees the disk at two different orientations due to a phenomenon called *gravitational lensing*. Accretion disks are powerful X-ray sources. Copyright © Legendary Pictures, Lynda Obst Productions, Paramount Pictures, Syncopy. Image credit: MovieStillsSB.com.

rial spiraling into bodies such as neutron stars and black holes, are powerful astrophysical sources of X-rays. In the film *Interstellar* (2014), the supermassive black hole Gartantua was depicted with an accretion disk heated to high enough temperatures to provide heat and light to planets orbiting. The planets, particularly the first planet the crew of *Endurance* visit, would be bathed in titanic amounts of highly energetic X-rays.

In the 1963 Roger Corman feature *The Man with the X-Ray Eyes*, Ray Milland—who starred in several horror and science fiction films in the 1960s and 1970s (including the role of Sire Uri in the original *Battlestar Galactica*)—appears as Dr. Xavier: a scientist who develops eye drops that allow him to see into the UV and X-ray portions of the EM spectrum. Though its cinematic artistry may not place it alongside movies such as *The Green Mile* (1999) or *The Color Purple* (1985), or even a web series like *Orange is the New Black* (2013–), it is still uncommon when a segment of the EM spectrum gets top billing in a movie title.

At the very high-energy end of the EM spectrum, 0.01 nm and shorter (30 EHz and above) is a form of radiation we have encountered previously: gamma rays. Gamma radiation is generated by radioactive decay processes, thermonuclear explosions, matter/antimatter pair annihilation, the nuclear fusion processes that generate power within the cores of stars, even supernovae. This form of radiation can be very damaging to biological tissues, and can also be mutagenic—these are the very same gamma rays that mutated the spider that bit Peter Parker[43], allowing him to become Spider-man, and that mutated Bruce Banner to become the Incredible Hulk.

In *Hulk* (2003), after a lab accident, Bruce Banner (Eric Bana) throws himself in front of an exposed gamma ray source to protect his colleague Harper (Kevin Rankin) from receiving a lethal exposure. Given the energy and penetrating power of gamma rays, and the damage they can do to living tissue, how would that scenario more have likely played out? Dr. Banner would have received a lethal dose of gamma rays, genetic enhancements notwithstanding. Harper would still have received a lethal dose. The wall behind Harper which, presumably, was shielded, would have received very nearly the same gamma ray exposure had neither Banner nor Harper been in the way.

X-Rays have nothing over gamma rays when it comes to film titles. One of the most unusual film titles in cinematic history was actually the name of the science fair project of one of the characters in the film, the 1972 Paul Newman film *The Effect of Gamma Rays on Man-in-the-Moon Marigolds*[44].

6.3 Ionizing vs. Nonionizing Radiation and Radiation Sickness

Radiation is not inherently bad, but radiation sickness is. The good news is that radiation sickness is caused by certain types of radiation only, radiation to which very few people are ever exposed. The bad news is that if you are exposed to it, you are in for a rough ride.

Radiation, particulate or electromagnetic, can be considered either *ionizing* or *nonionizing* depending upon how much energy it has. Radiation that has enough energy to strip electrons from atoms is called ionizing radiation. The three most common types of radioactive decay products—alpha, beta, and gamma radiation—can be forms of ionizing radiation. Some forms of radioactive decay emit positrons. Positrons are not inherently ionizing, but they are a form of antimatter, and it does not take them long to find an electron to rear-end and annihilate, thus creating gamma rays which *are* ionizing. It is these gamma rays that are detected during the PET scans often used to detect tumors in cancer treatment (PET stands for Positron Emission Tomography).

There are other radioisotopes that emit neutrons as they decay. Because neutrons do not possess an electrostatic charge, they are not repelled by electrons or protons, so neutron radiation penetrates very deeply into normal matter. Although the mechanisms are very different, high-energy neutron and gamma radiation share the property that they can split atomic nuclei. Al-

---

[43] In some, but not all, incarnations of Spider-Man.
[44] SPOILER: The rays are mutagenic in this film, too.

Fig. 6.7 Geiger counters on display at the National Atomic Testing Museum in Las Vegas. Some of these counters go back to the earliest days of atomic bomb testing in the Nevada desert. (Photograph by Stephen Cass)

though it would be rare for a neutron to strip electrons from an atom directly, they can attach themselves to an atomic nucleus, creating an unstable isotope that then decays. For this reason, neutron radiation is indirectly ionizing—the decay products are often ionizing.

The Geiger counter, a fixture in many a science fiction movie (when radiation was the true silent adversary), senses ionizing radiation. The main element of a Geiger counter is a noble gas like helium or neon in a tube that has a voltage applied across it. When atoms of the gas are ionized by radiation, the tube becomes temporarily conducting, registering a "click." More radiation means a higher rate of clicks. As ionizing radiation can be highly localized—the difference between an area with high radiation and low radiation can be a matter of centimeters—mobile Geiger counters are vital instruments when moving around places that have suffered widespread contamination, such are Chernobyl or Fukushima (Fig. 6.7).

Ionizing radiation can do severe damage to living tissue, resulting in radiation sickness, also known as Acute Radiation Syndrome (ARS). ARS results from the cumulative effect of the various forms of damage radiation on living tissue. The process of ionization, stripping electrons from atoms, releases energy, heats surrounding tissue, and can lead to redness, tenderness, swelling, blistering, burns and even open sores and bleeding for exposed skin. Extremely high-energy radiation, particularly gamma and neutron radiation, can turn stable atoms into unstable radioisotopes, releasing even more radiation and energy, thus causing even more damage.

Radiation can alter chemical bonds, and can turn water into toxic hydrogen peroxide, or create free radicals within a living organism. Free radicals are highly reactive atoms or molecules that readily, even aggressively, form chemical bonds with other elements and molecules, forming numerous types of compounds, many of which can be toxic. Living organisms require a small amount of free radicals to perform basic cellular functions, but when the concentration of free radicals becomes elevated beyond a body's ability to regulate it, it can be bad news for the organism.

Ionizing radiation can damage the cells lining the interior of the digestive system. Not only does this prevent them from functioning properly, it can also permit bacteria to migrate from their home in the digestive system, where bacteria are necessary for proper digestive function, into the blood as infection. Radiation can destroy white blood cells, leading to infections, red blood cells leading to anemia, and platelets leading to bleeding. Radiation can also damage soft tissues resulting in bleeding from the gums, mouth, and nose, and internally manifesting as bruising.

ARS victims can experience fever, nausea, vomiting (including blood), diarrhea and dehydration, bloody stool, and rectal bleeding, as well as ulcers throughout the entire digestive tract. Radiation can also cause hair loss.

Comparatively small doses result in the gastrointestinal effects listed, as well as bleeding and infections from reduced white cell count. Comparatively larger doses can result in neurological effects and rapid death. In the 2002 *Stargate SG-1* episode "Meridian," main character Dr. Daniel Jackson (Michael Shanks) is exposed to a lethal dose of radiation from an experiment gone awry. With heartbreaking clinical detachment, as well as impressive scientific accuracy, he describes to Colonel O'Neill (Richard Dean Anderson) the progression of symptoms that will culminate in his death:

```
                  DANIEL
The nausea will be followed by
tremors, convulsions, and
something called ataxia. Surface
tissue, brain tissue, and internal
organs will inflame and degrade, I
believe that's called necrosis.
And, based on the dose of
radiation I got, all that will
happen in the next 10 to 15 hours.
And if I don't drown in my own
blood and fluids first, I will
bleed to death, and there is no
medical treatment to prevent that.
```

The time to onset as well, as type of symptoms, depends upon the radiation dose, and from Daniel Jackson's description, with a life expectancy of 10–15 hour, we can infer that the dose of radiation he received was extremely high. Treatment of acute radiation syndrome involves blood transfusions and administering antibiotics, with bone marrow transplants in instances of high, but treatable, doses.

ARS can also be the "gift" that keeps on giving, with symptoms emerging years later. After the first round of signs and symptoms, a person with radiation sickness may have a period with no apparent illness, followed by the onset of new, more serious symptoms called chronic radiation syndrome (CRS). Not only can ionizing radiation damage living tissue, it can damage, destroy, or alter the DNA within the cells of living organisms. Since DNA is both the genetic code that tells cells how to divide, as well as how to produce proteins and other chemicals necessary for life, altering DNA can have profound ramifications. Normally a healthy organism can either repair damaged DNA or destroy the cell containing it through a process known as apoptosis. With ARS, the situation can arise where the DNA is neither repaired, nor does the cell send out a chemical signal telling the body that it is damaged. In this instance, the DNA replicates, passing on the damage to subsequent cell divisions. This can lead to various forms of cancers.

6.4 Why Is Radiation so Frightening?

People have seen the photographs and read the accounts of the survivors of Hiroshima and Nagasaki. People have seen the tests at Trinity, Bikini, and Eniwetok, and watched the evacuations of the towns surrounding Chernobyl and Fukushima.

Hollywood has helped as well. In addition to alleging that radiation could be the inciting incident behind the creation of every Kaiju monster, every scary mutation, and even beneficial mutations like those behind the *X-Men*, the *Fantastic Four*, and the *Hulk*, there have been more realistic screenplays that have shocked and terrified the public. *The Day After* in 1983 was a made-for-television film. It purported to chronicle the lives of a handful of people living in a small town in Kansas after a large-scale nuclear exchange between the United States and the Soviet Union. Its first-run broadcast was viewed by over 100 million people, making *The Day After* the highest-rated television film in history. The British equivalent, *Threads*, aired ten months later in 1984 and was, perhaps, even more terrifying and true-to-life.

> The release of atomic energy has not created a new problem. It has merely made more urgent the necessity of solving an existing one.
>
> Albert Einstein, Physicist

On the reimagined *Battlestar Galactica*, Executive Producer Ronald D. Moore eschewed more exotic forms of weaponry for the antagonist Cylons, and, instead, portrayed the scariest weapons he could think of—the garden variety thermonuclear warhead: "In the *Galactica* miniseries, when the Cylons attack the colonists, they attack them with thermonuclear weapons. They don't attack them with lasers and photon torpedoes, and strange things that don't exist. When you see a planet nuked, and you see those mushroom clouds, and hear about the destruction of entire cities by nuclear weapons, that is a much more terrifying and frightening idea than if you're saying fifteen thousand photon torpedoes were launched at Caprica. One is real and one is not."[45]

Radiation is also insidious. You can't see it, you can't smell it, and you can't taste it. Apart from sunlight, most people will never be exposed to the type of harsh radiation that can cause radiation sickness or delayed-onset maladies like cancer. Pepperdine University neuroscientist Jessica Cail explains why radiation is frightening, yet other behaviors that are far more dangerous, are not:

> You stand a better chance of dying on I-405 than you doing from some radiation cloud floating across from Fukushima, but yet we're not really concerned about jumping in the car—and some people don't wear their seatbelt, and they text, and eat, and put on make up while they're driving—and that's way more statistically dangerous. Studies do show that if you're in control of the situation, you feel much less afraid of it. The irony is that you have control your cigarettes, and then you don't take control of them and end up dying from that. But you have no control over this wave of radiation coming over the ocean. So it's that much more scary, but much less deadly. Usually it's harder to learn association between two things when there is a long delay between the cause and effect in the pairing. So cigarettes and poor diet and lack of exercise will not only kill you, but take a long time to kill you. Radiation, can kill immediately so it's just much more salient. You've only to see a few bad movies to imagine that.

6.5 So Where Did we Get the Notion of Giant Bugs and Reptiles?

Just like radiation in space can "flip" bits in spacecraft computers from 0 to 1 or vice-versa, a process known as a *single event upset*, so can radiation "flip" bits in a strand of DNA. Spontaneous modifications to the genetic code of an or-

---

[45] Interview on the BBC *Cult TV* 20 Feb. 2004.

ganism is known as a *mutation*. Any substance or process, like radiation, chemicals, entropy, that results in mutations is called mutagenic—radiation is both carcinogenic and mutagenic. While mutation is an essential phenomenon that drives evolution, most mutations are harmful—or, at the very least, benign evolutionary dead ends. Some mutations can have very profound effects.

> There's almost no food that isn't genetically modified. Genetic modification is the basis of all evolution. Things change because our planet is subjected to a lot of radiation, which causes DNA damage, which gets repaired, but results in mutations....
>
> Nina Fedoroff, Biologist, Syracuse University

If the cellular mutation happens within the reproductive system of an organism, within sperm or ova for example, mutations can be passed along to decendants. This is the raw material of evolution (later enhanced by the genome-shuffling invention of sex). It is an important point, and a line that is frequently blurred by Hollywood, but while there are many ill effects that radiation can visit upon an individual, the kind of mutation that can, say, radically alter an organism's body plan, happens to the *descendents* of the radiation-exposed person or animal or plant only (mutations *can* occur within the cells of an existing body, but the result is a tumor, not superpowers).

Unfortunately mutations passed to descendants do not necessarily confer any advantages either, having either no effect[46] or even manifesting as deformities such as polydactylism, extra or misplaced limbs, missing organs, malformed limbs, extra small head size (microcephalism), extra large head size (macrocephalism), and after that the list gets nasty. Sotos Syndrome and Weaver Syndrome are two forms of mutation-induced gigantism—where an organism is normal in all respects, but its size. On rare occasion, radiogenic mutations can manifest as forms of gigantism.

What is unclear is why gigantism occurred so often in science fiction of the 1950s and early 1960s, with its peak around 1957. While gigantism has been reported in plant species in the radioactive region surrounding the Chernobyl nuclear reactor, the Radiation Effects Research Foundation, a joint U.S./Japan research organization, reported, "No statistically significant increase in major birth defects or other untoward pregnancy outcomes was seen among

---

[46] Benign mutations can accumulate in noncoding regions of the genome that no longer have any direct biological impact on the organism. Knowing the rate at which these mutations accumulate in DNA has allowed geneticists to construct molecular clocks that have allowed us to measure when two species have diverged from a common ancestor, and even estimate when humans began migrating out of Africa.

Fig. 6.8 Godzilla, or Gojira, from *Godzilla* (1954). (Copyright © Toho Film (Eiga) Co. Ltd. Image credit: MovieStillsDB.com)

children of survivors. Monitoring of nearly all pregnancies in Hiroshima and Nagasaki began in 1948 and continued for six years." So not only were the children of the Hiroshima and Nagasaki victims not giants, research reveals that mutation is not a big part of that equation, though research is ongoing.

So where did we get radioactive giants? Giant monster films began long before World War II, with *The Lost World* (1925) and *King King* (1933), but they really reached their peak after World War II, when atomic tests regularly lit up the skies of places like Nevada and the Bikini atoll. With spectacular explosion filling newsreels and the poorly understood nature of radiation among viewers, it wasn't long before a screenwriter said to himself, "I want to sell a 300-foot-tall behemoth with 'atomic breath.' Clearly, radiation." The result was things like the giant ants of 1954's *Them* and the eponymous spider of 1955's *Tarantula*. It was probably in Japan, though, where the fear of radiation had seeped deeply into the national subconscious, that the radioactive-giant genre reached its apotheosis in *Godzilla*, released just a few month after *Them* (Fig. 6.8). Eventually, long after home-grown giant radioactive bugs had fallen out of favor in the United States, the popularity of *Godzilla* and its Kaiju descendants (on both the big screen and small), would lead to a Western revival of the form, with two American versions of *Godzilla* released in 1998 and 2014, as well as the somewhat more critically successful 2013 *Pacific Rim*[47].

---

[47] *Pacific Rim* was a big deal, with wide-ranging implications in the film industry. It was a very rare case, these days, of an original film—one that was not a sequel, a book property, or a reboot—that made a lot of money.

Science Box: How Hot Is the Human Torch?

How hot is Johnny Storm[48], aka "The Human Torch," of the Fantastic Four Fig. 6B.1? We can not say for certain, but we can apply reasonable constraints based upon observation[49]. Storm's superpower is that he can surround his body in super-hot plasma without harm. He has learned to manipulate the plasma envelope to give him the ability to fly.

Fig. 6B.1 Chris Evans is Johnny Storm, aka the Human Torch, in *The Fantastic Four* (2005). Copyright © Twentieth Century Fox. Image credit: MovieStillsDB.com.

In *The Fantastic Four* (2005), Dr. Doom fires a heat-seeking missile at torchified Johnny Storm, whose skillful flying allows him to evade it. Now this is an observation that might help us. Heat-seeking, or IR-seeking, missiles operate in the mid-IR range of the EM spectrum, from 8000 nm (37.5 THz) to 3000 nm (100 THz). Using Wien's Law,

$$\lambda_{max} = \frac{2,900,000}{T(\text{K})} \text{nm},$$

and solving for temperature, T, yields,

$$T(\text{K}) = \frac{2,900,000}{\lambda_{max}(\text{nm})}.$$

If we solve this for the longer wavelength end of the mid-IR, we get

$$T(\text{K}) = \frac{2,900,000}{8000 \text{ nm}} = 362 \text{ K},$$

---

[48] Chris Evans fans just said, "Incredibly." Wipe that drool off your chin.
[49] Chris Evans fans just said, "Now that's what I'm talking about!"

so T is 362 K or 89 °C (192° F). That's not very hot, not even the temperature of boiling water. At the more energetic end of the range,

$$T(K) = \frac{2,900,000}{8000 \text{ nm}} = 967 \text{ K},$$

which corresponds to 694 °C (1281 °F). That doesn't seem very hot either. The temperature of a burning match ranges between 600 °C (1112 °F) and 800 °C (1470 °F). Johnny Storm is not "The Human Match." If he was, he would certainly be no match for supervillains the likes of Dr. Doom[50]. A blowtorch burns around 2000 °C (3630 °F); an acetylene welding torch burns around 3500 °C (6332 °F). Now that is more like it! The latter temperature is hotter than the surface of small stars.

Perhaps we are going about this wrong, and focusing on the wrong observation. In both films and comic books, Torch's color is always depicted as a reddish-orange. The middle of the range of colors that can be called "red" is 650 nm, corresponding to a temperature of 4461 K (4188 °C or 7570 °F); the mid range for orange is 590 nm or 4915 K (4642 °C or 8387 °F). So the plasma enveloping Johnny Storm is probably in the 4700–4800 Kelvin range. He's never depicted as yellow[51], or green, or blue, so that gives us an upper bound.

If Storm's temperature is out of the operational range of a heat-seeking missile, is it scientifically plausible that Dr. Doom's missile would lock on to Johnny Storm?[52] A feature clear in Fig. 6A.1 is that the thermal energy curve of hotter objects are peaked at higher energies, they also emit more radiation at all wavelengths that cooler objects. This is why fighter craft with heat-seeking missiles on their tails deploy decoy flares; many examples can be found in the reimagined Battlestar Galactica and Airwolf (1984–1986, 1987). Anti-aircraft decoy flares burn between 2000 K to about 3000 K, far hotter than aircraft tailpipes, but they emit more IR radiation at all wavelengths than a hot tailpipe. At a temperature hotter than the surface of most stars, an anti-aircraft heat-seeking missile would find The Human Torch a tempting target[53].

---

[50] If he was, would the other three members be similarly demoted: Mr. Above Average, The Translucent Woman, and Rocky?

[51] We will ignore Storm's nova flame, which is much hotter. We're just interested in his quiescent state.

[52] For perspective, let's remember that we are talking about a man who can fly while enveloped in superheated plasma.

[53] Chris Evans fans just said, "So Say We All!"

7

A Quantum of Weirdness

I think I can safely say that nobody understands quantum mechanics.
Richard Feynman, Physicist

When you open your mind to the impossible, sometimes you find the truth.
Walter Bishop, "Olivia. In the lab. With The Revolver", *Fringe*

I bet any quantum mechanic in the service would give the rest of his life to fool around with this gadget.
Chief Engineer Quinn, *Forbidden Planet*

Quantum mechanics is *weird*. It is a realm where matter can teleport from one place to another, the building blocks of solid matter can be as unsubstantial as a ghost, and the emptiest void of space boils with activity. Reality itself is a fickle creature, with randomness at the very heart of existence. No wonder, then, that screenwriters have often latched onto quantum mechanics as the ultimate get-out-of-jail-free card. Can't figure out even a remotely plausible explanation for some incredible piece of machinery? Just have one of the characters utter some technobabble heavily laced with "quantum this" or "quantum that."

Although this has always annoyed physicists, for much of screen science-fiction's history it is difficult to argue that writers should have been digging deep into the interconnecting oddities of quantum theory. Sure, quantum effects are essential to understanding how technologies like lasers or transistors work, but it is not like the engineer writing firmware for a microprocessor inside a Blu-ray player needs to be familiar with the Schrödinger wave equation, or any of the other equations at the heart of quantum mechanics[1]. For the

---

[1] For the record, the Schrödinger wave equation is: $i\hbar\dfrac{\partial}{\partial t}\Psi = \hat{H}\Psi$. It describes how the state of a quantum system, such as an electron orbiting the nucleus of an atom, changes over time. The fact that this is the equation for a *wave*, and yet is used to describe the behavior of *particles* like electrons, is at the heart of much of the counter-intuitive aspects of quantum mechanics.

most part, throughout the 20th century, the really outlandish implications of quantum mechanics—like how a particle can exist in two places at once—was tucked away in everyday reality's basement, only obviously manifesting in carefully constructed laboratory experiments.

For most scientists, the world was nicely divided into the microscopic (in this case meaning subatomic scales) and the macroscopic (everything else). In the subatomic world, quantum mechanics rules. In the macroscopic world, good old-fashioned "classical" physics holds sway. Classical physics is most of the physics that you might have been taught in high school, i.e., calculating how high a ball might go if thrown upwards with a given velocity. In classical physics, particles can be treated as tiny billiard balls, and their position and speed determined with theoretically perfect precision. If classical physics was good enough to enable rocket scientists to drop probes onto the surface of other planets tens of millions of miles away with pinpoint accuracy, surely it was good enough for the average screenplay.

Things have been changing in recent years. Scientists and engineers have been increasingly probing the *mesoscopic* realm, the strange borderlands between the microscopic and macroscopic. The result has been that technologies are emerging—such as quantum computing, quantum cryptography, and tools that can image and manipulate matter on the atomic scale—that bring quantum weirdness up close and personal. Evidence is also mounting that many critical biological processes—such as plant photosynthesis and animal vision—are more reliant on quantum effects than previously suspected (some scientists, like Roger Penrose, have even proposed that human consciousness relies directly on quantum processes, but the general run of opinion is against this particular idea.)

That said, there has long been a particular sub-genre of shows and movies that have relied heavily on quantum effects for more than just technobabble. Any plot that features a parallel universe similar to our own, while differing in such matters as how evil, brave, bisexual, or wealthy various characters are, is playing out the Many Worlds interpretation of quantum mechanics, proposed by Hugh Everett in 1957. The many-worlds theory is an attempt to solve a very deep problem in quantum mechanics, with implications for all the other quantum technologies we'll be discussing in both this chapter, and in Chapter Ten—"Moving in Stereo: Parallel Universes."

Everything that can happen, happens. It has to end well, and it has to end badly. It has to end every way it can.

Will Graham, *Hannibal*, "Primavera"

7.1 Catching a Wave

In quantum theory, a physical system evolves according to the aforementioned Schrödinger wave equation. So, say you're looking to model how an electron behaves while orbiting an atom. Plug key values into the Schrödinger equation, such as the charge and energy of the electron, and the result is a *wave function*. This wave function *doesn't* pinpoint the electron's position at any given moment, rather it describes the *probability* of finding an electron at any given location in the space around the nucleus. For some regions of space, the wave function will be vanishingly small, meaning the electron will, practically speaking, never be found there. In other regions, the probability of finding the electron there will be so high that, for most purposes, the electron can be considered to be confined to those regions. This is a quantum mechanical way of saying that electrons can occupy only fixed orbits, which we discussed in the previous chapter.

> That's preposterous! Your Honor, the Fermi-Dirac function is, for any system of identical Fermions in equilibrium, the probability that a quantum state of energy E is occupied. My word, man! Don't you know your quantum statistics?
>
> Brain, *Pinky and the Brain,* "Of Mouse and Man"

Now here's the weird bit—it's the act of *finding* the electron that determines where exactly it is located, turning probability into position. This is totally at odds with how things work in the macroscopic world. Imagine two macroscopic boxes. We know the boxes contain a prize, and in this case we also know there's a 60 % chance of it being in one box and a 40 % chance of it being in the other. When you open a box to look inside, it's not the case that *that* is the moment the prize decides which box it's in. It was always in one box or the other, even if *you* didn't know which. In the quantum world, until you look inside a box, the prize's existence is spread over *both* boxes, (although in this example, it's existence is, in effect, a little more in one box than the other) waiting for you to open one. Only then will it decide in which box it is located.

If you're finding this difficult to accept, you're in good company. Albert Einstein never fully accepted this view of reality, saying "I am convinced that [God] does not play dice." By this he meant that he believed an electron's existence was always perfectly defined by so-called *hidden variables* that were not a part of quantum theory. Experiments conducted in the 1970s and 1980s proved that Einstein's hidden variables did not exist (we will revisit these experiments later, as they proved to be the foundation of quantum cryptography and teleportation).[2]

Anyone who is not shocked by quantum theory has not understood it.

Neils Bohr, Physicist

The technical term for finding an electron and making it "pick" a location is *collapsing its wave function*. Every time any quantum system's probability wave is observed, it collapses and the electron picks just one of the possibilities, or states, permitted by the wave function. This introduces a knotty problem into quantum theory. Just what exactly defines an observation? Who, or what, is doing the observing?

This problem is at the heart of the *Copenhagen Interpretation* of quantum mechanics, developed in the 1920s. In the Copenhagen interpretation, the wave function and all the other equations in quantum mechanics are more-or-less just handy mathematical tools that can be used to predict the outcome of experiments. In a nutshell, the Copenhagen interpretation says: "Don't worry about what is exactly meant by wave functions or observations, they'll take of themselves, just do the math."

This pragmatic approach of sidestepping defining an observer, or how one causes a wave function to collapse, has never sat well with more philosophically-minded physicists. Schrödinger himself proposed his famous "Schrödinger's Cat" thought experiment as a way of illustrating his objections to the Copenhagen interpretation.

In this thought experiment, a cat is sealed in a box along with a lump of some mildly radioactive substance; a Geiger counter equipped with a hammer; and a vial of poison gas. Over the period of, say, one hour, there is a 50/50 chance, as given by a quantum probability wave function, that one of the atoms in the radioactive substance will decay and cause the Geiger counter to register a "click" of radiation. If the counter clicks, the hammer will drop, smashing open the vial, killing the cat.

At the end of the hour, before the box is opened, what is the state of the cat? Is it alive? Is it dead? From a quantum mechanical standpoint, until the cat is observed, it exists in a state of quantum suspension, both alive and dead. This raises the thorny question: does the cat qualify as an observer? Assuming the poison acts instantly, can a *dead* cat be an observer? In other words, when

---

[2] In a twist of scientific irony, although Albert Einstein was not a fan of quantum mechanics, he won the Nobel Prize for explaining the Photoelectric Effect—a quantum mechanical phenomenon. God may not play dice, but he does, apparently, love irony.

exactly is the cat's fate determined? The concept of quantum suspension—where two states of being co-exist—is technically known as *superposition* of states. Superposition is an important idea, which, as we'll see, plays a key role in quantum computing and other quantum technologies.

> The world is not made up of particles and waves and beams of light with a definite existence. Instead, the world works in a much more exploratory way. It is aware of all the possibilities at once and trying them out all the time. That is a hard thing to picture.
>
> Neil Turok, Physicist

7.2 The Age of Uncertainty

Another phenomenon that plays a key role in quantum technologies, and which is a pillar of quantum mechanics, is the *Heisenberg Uncertainty Principle*. Heisenberg's uncertainty principle puts a limit on how accurately we can know properties of the particles that make up our universe. Specifically, there are pairs of properties, known as *complementary variables*—such as position and momentum—in which more accurate knowledge of one property increases our ignorance of the other. In other words, the more precisely you measure the position of a particle, the more uncertain its momentum becomes, and *vice versa*. The Heisenberg uncertainty principle explains why a temperature of absolute zero is not the point at which atomic motion stops. If atoms stopped, both their positions and velocities would be knowable.

At this point, we need to undo some damage regarding a common misconception of what the Heisenberg uncertainty principle actually means, blame for which can actually be laid at the door of many a science textbook rather than any science fiction screenplay. The uncertainty principle is often confused with the *observer effect*. In the observer effect, the act of observation is so intrusive that it changes the thing being observed.

Concerns about observer effects are common throughout science—a tremendous amount of effort has gone into developing nonintrusive medical imaging techniques, such as Magnetic Resonance Imaging (MRI), so that we can, for example, observe how the brain works without cutting a patient's head open and possibly damaging the very organ under investigation.

In another example, in anthropology, modern scientists often use the technique of *participant observation* to minimize the changes that occur in a society when even well-meaning observers turn up and make people feel like rats in a laboratory maze. With participant observation, anthropologists try to engage in the day-to-day life of the society as much as possible, without trying to do things the "proper" way they've been taught in their home society. (In the *Star Trek* franchise, this means that in order to avoid altering the destinies of pre-warp civilizations, observations are made from hidden bases, as in the 1989 *Star Trek: The Next Generation* episode "Who Watches the Watchers?" or undercover operations are performed in which Starfleet personnel are disguised as natives, as in 1991's *Next Generation* episode "First Contact").

In physics, observer effects arise because most methods of measuring parameters of a physical object require interacting with it in some way. To determine the location of a particle, for example, typically requires bouncing photons off it. This is no big deal for large objects, but at the quantum level the energy of a single probing photon can alter a particle's trajectory, even change its energy. Because higher energy photons provide better spatial resolution due to having smaller wavelengths, the more and more precisely you want to fix the location of a particle, the more and more you alter its trajectory, thus increasing the uncertainty in its location the following moment. It is this observer effect that has often been confused, even in physics courses and textbooks, with the actual Heisenberg uncertainty principle, which goes much deeper and—say it with us—*weirder*.

It is possible, in some special cases, to measure the position of a particle *without* physically interacting with it. (How this is accomplished would take too long to explain here, so you'll have to trust us, but you can read about it in some of the books we recommend at the end of this one.) Here's the weird thing—even *without* interacting with the particle, the Heisenberg uncertainty principle means that learning something about its position still increases uncertainty regarding its momentum. To use a horribly anthropocentric analogy, it is as if the particle somehow "knows" its location is being spied upon, and like a secret agent who realizes her car is being followed, suddenly tries to mix up her speed, in an attempt to lose the pursuers.

The Heisenberg uncertainty principle has another sting in its tail. It does not merely dictate that measuring one complementary variable increases the uncertainty in the other variable. If that was the only implication, then one might be happy to know the position of the particle exactly for the price of absolute ignorance in the momentum. That would be too easy. The uncertainty principle states that the *combined* uncertainty has to be at least a value called *Planck's constant* divided by 4 times π. It is, therefore, impossible, even in principle, to exactly know the position or momentum of a particle. There's a fundamental fuzziness to reality, with the fuzziness determined by Planck's constant. (Incidentally, Planck's constant is represented by the letter h, and, for convenience, $h/2\pi$ is written as $\hbar$ (called h-bar). If you flick back to the start of this chapter and look at the footnote showing Schrödinger's wave equation, you'll spot $\hbar$ lurking within.)

Planck's constant is an insanely small number, which is why we don't notice the universe's fuzziness in classical physics. It still has some mind-boggling implications, though. One of these is that *there is no such thing as empty space.* To see why this is so, let's switch from position and momentum to another pair of complementary variables: energy and time[3]. What we've said above for position and momentum applies to energy and time: it is impossible to know *exactly* the energy of a quantum system, but instead there must be at least a teeny bit of uncertainty, as dictated by the Planck constant.

> To let understanding stop at what cannot be understood is a high attainment. Those who cannot do it will be destroyed on the lathe of heaven.
>
> Ursula K. LeGuin, *The Lathe of Heaven* (1971)

Consider a patch of outer space, say a cubic meter somewhere in the void between galaxies. It sure looks empty—rays of light and even the occasional stray atom will zip through it unimpeded. According the Einstein's famous equation, $E = mc^2$, there is an equivalence between mass and energy. So to say a volume of space has *absolutely* nothing in it, is equivalent to saying its energy is *exactly* zero which, as we've just seen, is forbidden by the uncertainty principle. So that space has to have *something* in it. At the quantum level, the concept of "empty" has no meaning.

What fills the space is so-called *quantum foam*. Subatomic particles—dubbed *virtual particles*—are popping in and out of existence all the time. The reason they don't stick around to fill up all space is because energy's complementary variable is *time*. A particle can exist only within a brief time window of uncertainty that's inversely proportional to the virtual particle's mass. (Virtual particles are actually created in matter and antimatter pairs, and the energy released when they mutually annihilate each other a moment later "pays back" the energy used to create them. A special situation is believed to occur however, right at the event horizon of a black hole, where one of the virtual particles crosses over the event horizon and gets swallowed up, but the other particle escapes. What pays for this particle's energy? The answer is: the black hole, which shrinks a tiny amount in accordance with the mass of the escaping particle. This means that black holes don't exist forever, as previously thought. The escaping virtual particles—now considered real particles—are what make up *Hawking radiation*. Stephen Hawking came up

---

[3] Explaining why two particular variables end up forming a complementary pair is beyond the scope of this book, but if you mutter things like "orthogonal state vectors" and "phase space," you can fake knowing.

with this idea in 1974, and it's one of the reasons why he's such a big kahuna in theoretical physics.).

That virtual particles actually do exist can be proved directly in the lab by measuring the attraction between two closely spaced metal plates due to a phenomenon known as the *Casimir Effect*. (For the details of how this effect works, and how it might be used to stabilize a wormhole, see Chapter Nine: "Shortcuts through Time and Space"). The existence of virtual particles, or quantum foam, also goes a long way towards explaining why the two foundations of theoretical physics—quantum mechanics and relativity—do not play nicely together. Quantum mechanics describes the behavior of the very tiny, and our understanding of it is the basis for technologies like semiconductors, solar panels, and even television. Relativity describes the macroscopic universe: objects that are very massive or things moving very fast. Our understanding of relativity is a necessary component of technologies like satellite communications and GPS. At some scale, though, these two descriptions of the Universe—each of which works very well in their individual realms—simply do not jibe.

The main reason they don't jibe is quantum foam. According to Einstein's theory of General Relativity (more on this in Chapter Nine: "Shortcuts through Time and Space"), gravitation is less of an invisible force of attraction between two objects, and is described better by the notion that any object with mass warps space. The more mass, the greater the warp. Because virtual particles do possess mass, at quantum size scales, gravitation is ever shifting. Gravitation can't be represented by a smooth warping of space, then, but by a constantly-changing frothy landscape. Attempts to build a consistent quantum theory of gravity have so far failed, although there are several contenders such as *loop quantum gravity* and whatever the latest flavor of string theory is.

The fact that even an empty vacuum does possess some mass/energy has inspired some inventors to try and find ways to harness this so-called *zero-point energy* as an unlimited power source. To date, all such attempts have failed, and scientists are skeptical that the idea can ever work in practice[4]. Still, it has proved to be a handy tool for science fiction writers, most notably in the *Stargate* franchise. In *Stargate: SG-1* (1997–2007) Zero Point Modules, or ZPMs, are introduced as an energy source developed by the same Ancients who originally built the stargate wormhole network that our heroes use to explore the galaxy. ZPMs can output enough energy to support extremely heavy loads, such as powering a planetary defense system or sustaining a wormhole to another galaxy. A minor element in the original *SG-1* series, ZPM treasure hunts across the Pegasus galaxy became something of a theme for the spin-off series *Stargate Atlantis* (2004–2009) as the city-sized spaceship that is Atlantis required at least one, and ideally three, ZPMs for full operation of all its capabilities.

---

[4] This is simply the modern manifestation of the perpetual motion machine.

Stargate's writers were smart enough to give their ZPMs finite lifetimes. Though this was not a technological limitation born of true zero-point physics, it did enable important sources of narrative tension, as otherwise the ZPMs would have made life too easy for the heroes in *Atlantis*.

The energy/time version of the uncertainty principle also throws cold water on another popular device in science fiction—the stasis field. Put somebody or something in a stasis field and time within the field comes to a halt, at least this is the intent in the narrative. Claiming that no time passes within the stasis field runs afoul of the same problem of saying that no energy exists in a vacuum.

Dave Lister winds up stranded in the far future in the sci-fi comedy *Red Dwarf* (1988–) after he's put in a stasis pod, ostensibly for an 18 month punishment, on board the eponymous mining ship. While Lister is serving his punishment, the entire crew is killed by radiation that takes three million years to decay to safe levels. Apparently the writers were aware of the quantum issues on *Red Dwarf*, as the character of Frank Toddhunter tries to reassure Lister by describing the operation of the stasis pod aboard the ship with this bit of technobabble: "The stasis room creates a static field of time. See, just as X-rays can't pass through lead, time cannot penetrate a stasis field. So, although you exist, you no longer exist in time, and for you time itself does not exist. You see, although you're still a mass, you are no longer an event in spacetime, you are a non-event mass with a quantum probability of zero."

Some time would have to pass in a stasis field, but, to be fair, it could be at such an imperceptibly slow rate as to appear that time has halted, at least to human perception. This issue is why the writers of the 2011 "Reprise" episode of *Eureka*, which featured a stasis field, made sure the episode made explicit reference that time passed very slowly within the field, rather than being completely frozen.

The wave nature of Schrödinger's equation, along with the Heisenberg uncertainty principle and the concept of complementary variables, also gives rise to *wave-particle duality* in quantum mechanics. This means that every entity in the universe, depending upon how you view it, can be either a particle or wave: an object's "waveness" shrinks in proportion to how much its "particleness" is measured. For example, normally we treat light as a wave, and different colors are simply light waves of different energies. Treating light as a wave allows us to understand why the sky is blue, how conventional microscopes work, and how Internet traffic can be sent through kilometers of optical fibers. We can also treat a light beam as a stream of particles—photons—where the color of the photon corresponds to its energy, which is great for understanding how things like solar cells or digital cameras work.

Conversely, we can take a particle like an electron, and cause it to behave as a wave. This is the basis of *electron diffraction*, used to study crystalline structures on the atomic level. The electron waves interfere with each other in a way that's dependent upon how the atoms are spaced within the crystal.

Particles are waves; waves are particles. So often it is the case that Nature dreams up scenarios far more fantastic than any fictional narrative. Just when you believe Nature can get no weirder, you find yourself entangled in even more outlandish scenarios.

> Life is a wave, which in no two consecutive moments of its existence is composed of the same particles.
>
> John Tyndall, Physicist

7.3 Entangled: It Doesn't Happen to Just Rapunzel

If implications of the uncertainty principle seem weird, the implications of *quantum entanglement* are so strange that Albert Einstien famously referred to it as "Spooky action at a distance." Quantum entanglement is a consequence of the concept of a superposition discussed earlier—that prior to making a measurement of a quantum system, the system exists simultaneously in all possible states. Making a measurement forces the wave function to collapse into one definite state. Studying entanglement in the 1970s and 1980s demonstrated in the lab that Einstein's belief in hidden variables was wrong, and that it really does appear (from the point of a view of a single universe at least) that randomness is at the heart of reality, and that the act of measurement—whatever that means exactly—plays a key role in determining how that randomness plays out.

To get a handle on entanglement, we need to expand our definition of a quantum system. Until now, we've focused primarily on quantum systems of just one particle, typically an electron. It is possible to make more than one particle part of a single quantum system. Particles that are part of the same quantum state are said to be *entangled*. If you're careful to prevent the particle from interacting with anything else along the way, it's possible to separate two entangled particles by a large distance while maintaining the entanglement, keeping them together in a quantum superposition. If the particle does interact along the way, the superposition is likely to collapse before you want it to, a process known as *decoherence*. Say, however, you successfully manage to send one entangled particle to the Moon, and then perform a measurement on the other particle back on Earth. The measurement forces the Earth particle to pick a state as its wavefunction collapses—the entangled particle on the Moon will typically pick the complementary state, wherever it is. This effect happens *instantaneously*, wherever the other particle happens to be. Science currently has no real clue about how this effect works.

> Science is made up of so many things that appear obvious after they are explained.
>
> Pardo Kynes, *Dune,* Frank Herbert

With the right choice of quantum systems, once we measure what the state of the nearby particle is, we also know exactly the state of the distant particle, whatever it is. This long-distance entanglement effect has been demonstrated in a number of ways, but the easiest-to-understand setup uses polarized photons. The photons are created together, so they start off entangled and in a superposition of states with regard to their polarization: in this case they can either be circularly polarized to the right, or circularly polarized to the left. Don't worry if you're not familiar with the details of circular polarization. What's important to remember is that when the polarizations of the photons are measured, one *must* be left polarized and the other *must* be right polarized. You cannot have both particles left-polarized or both right-polarized.

Prior to any measurements that could force a state change, their entangled superposition of states means that each photon simultaneously exists as both left and right polarized. Again, send one photon to the Moon, and measure the polarization of the other photon when it arrives there (we could store the Earth photon in, say, a long loop of fiber-optic cable while the Moon photon is travelling). Say the Earth photon turns out to be left polarized. Instantly the photon on the Moon becomes right polarized. In fact, if we could manage to avoid decoherence, the other photon could be on the other side of the galaxy and it would instantly become right polarized at the moment of measurement.

If it sounds like this effect could be used as the basis for a faster-than-light communication system like the subspace communications used by Starfleet in *Star Trek*, there's bad news: you can't send information faster-than-light this way because there is no way to control which state the photon on Earth will choose—it will be random, so observing the other photon will just replicate a meaningless string of right and left polarizations. However, as we'll see, it can be handy in some slower-than-light technologies.

Entanglement can also be useful in cryptography to form the basis of an absolutely secure mode of communications[5]. There are a number of ways to

---

[5] Well, almost. Theoretically it's a perfect system, but theoretical ideas have to be implemented by imperfect devices, which can provide a teeny bit of wiggle room for a really determined attacker.

implement such a system, but the essence of one approach is to send entangled photons down a fiber-optic line and use the photons as the basis of a cryptographic key that's used to encrypt data. If an eavesdropper tries to peek at what's going over the fiber-optic line, she will cause the entangled superposition to collapse. This can be detected, and so both parties know not to trust that cryptographic key to send any information.

7.4 The Ultimate Computer

Superpositions can also be used to power an entirely new breed of computers: *quantum computers*. Quantum computers could be the key to the biggest leap in computing power since the invention of the integrated circuit, perhaps bigger. For some tasks, they will be able to perform in a moment what it would take a traditional computer years or centuries. They're becoming increasingly popular with screenwriters too: An optical quantum system (with chips composed of diamonds) was used to power the central computer of the labs at General Dynamics on *Eureka*, and when Johnny Depp's terminally ill character Will Craster has his consciousness digitized in 2014's *Transcendence*, it's sustained by an array of quantum processors.

Simple quantum computing systems have already been demonstrated in the lab, and one company, D-wave, has begun selling a commercial quantum computer (early customers include Google and Lockheed Martin), though just how quantum their machine really is, is a matter of some debate.

One of the neat tricks some types of quantum computer would able to perform (for technical reasons, D-Wave's system can't do this even if it *is* definitely doing quantum computing) is to factor large numbers. Factoring a number refers to the process of finding smaller numbers that can be multiplied together to make that number (ignoring fractions—we're only interested in integers here, that is, whole numbers). For example, the number 12 has the factors 12, 6, 3, 4, 2, and 1. The factors of the number 77 are 11 and 33, ignoring this time the trivial factors of the number itself and 1. Factoring numbers sounds incredibly boooooooring, until you realize that multiplying large prime numbers together is the basis of all modern cryptography.

It turns out that traditional computers have an incredibly hard time factoring large numbers, so it is reasonably easy to pick a number that would take the most powerful supercomputer in the world hundreds of thousands of years to factor, if not longer. So a computer that could factor large numbers quickly would instantly make virtually all the encrypted traffic on the Internet as easy to read as if someone gave you every last password. Such a device

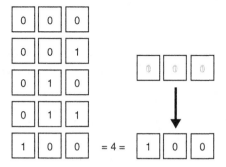

Fig. 7.1 A conventional computer memory must hold either a *1* or *0*. In many cases, in a conventional computer, to test if a particular number—in this case, 4—is the solution to a problem, every number must be tested until the right one is found. On the *left* we see the binary representations for the numbers, 0, 1, 2, 3, and 4 occupying a 3-bit memory in turn before the answer is found. In a quantum computer, *right*, each bit of the memory exists in a simultaneous superposition of 1 and 0s, which, when tested, allows the right number to be found in one go.

was actually the MacGuffin used to propel the plot of 1992's hacker movie *Sneakers*, in which the government is after a black box that can factor large numbers. Although it's never described as a quantum computer in the movie, if it walks like a duck and talks like a duck....

Quantum computers acquire their power by using *qubits* to perform calculations. In a conventional silicon computer, information is stored as bits, binary ones and zeros encoded as electrical charges in semiconductors. The computer processes these bits using elaborate arrangements of transistors. Only so many transistors fit on an integrated circuit chip, each of which has to be in a definite on or off, zero or one, state. So in order for a traditional computer to test the first billion numbers to determine if one of them is the factor of a large number, it must generate each number in turn, test it, then proceed to the next. Qubits can be forced into a superposition of states, so each qubit is both a one and zero at once. In a regular computer, chaining bits together to form *bytes* allows numbers larger than one to be stored (one 8-bit byte can store an integer as large as 256, while a 32-bit (four byte) integer can be as big as 4,294,967,296). Chaining qubits together in one elaborate superposition allows not only one large number to be stored *but also simultaneously every smaller number as well*. Essentially, they parallel process at a quantum level.

Simplifying (…a lot; the math is pretty hairy), it is then possible to collapse the superposition in such a way that the binary number that remains, after each qubit selects if it is going to be a one or a zero, is the number that is the factor you were looking for. In practice it's very difficult to maintain coherence between more than a handful of bits, while a typical encryption key is hundreds to thousands of bits long. Our secrets are safe … for now (Fig. 7.1).

7.5 Quantum Leaps

Superposition and entanglement also play key roles in *quantum teleportation*. Quantum teleportation isn't quite like the method of teleportation used in the *Star Trek* franchise. In *Star Trek,* matter can be beamed from a teleporter pad to any point in range (barring any narratively imposed shielding or other interference). Quantum teleportation is much more like the kind of teleportation seen in the movie *The Fly* (1958 and 1986 versions) or in the *Stargate* franchise. An object is placed in one chamber, teleportation goodness happens, and the object emerges from another chamber. Unfortunately for horror fans, even a fly is far too big and complex an object for current teleportation techniques, which so far have only managed to "teleport" individual atoms.

In fact, the term "teleport" is something of a misnomer in this context. What's going on in the chambers is actually a destructive form of one-shot replication. The object in the first chamber is scrambled beyond recognition by the teleportation process (which could explain some of the nasty teleporter malfunctions shown on screen. Both SAC and KRG remember their skin crawling as young 'uns following the radio message that *Enterprise* receives following such a malfunction in 1979's *Star Trek: The Motion Picture*[6]: "What we got back didn't live long. Fortunately.") What gets transmitted from one chamber to the other in quantum teleportation isn't matter, but *information*. Like stamping clay with a mold, the second chamber must contain all the actual matter that will be configured into the teleported entity.

The basis of quantum teleportation is that all particles can be completely described by a series of *quantum numbers*. Two particles that have the same quantum numbers are indistinguishable from one another.

Science Box: Miniaturization—Black Holes vs. Fantastic Voyage

If you're a Hollywood writer or showrunner for a science fiction series, here's an April Fool's prank to play on your science advisor, one that's guaranteed to elicit an uproarious response: say that you are going to do an episode where your stars are going to be miniaturized. Want to twist the screw even more? Say that it'll be the advisor's job to "sell" it. It is your choice how long you let your advisor sweat, moan, hyperventilate, curse, and throw things before you letting him or her off the hook with "April Fools." Expect a "You are *such* an asshole!" in reply.

When André Bormanis, science advisor of *Star Trek: Deep Space Nine*, first received the script for "One Little Ship," a story where a spacecraft, a runabout, and its crew are miniaturized, his initial response was, "You just gave me the one plot I'd hoped I'd never see."

Miniaturization has been something of a recurring theme in science fiction. There was the *Fantastic Voyage* (1966) (which also spawned a 1968 television series and a current remake effort), *The Incredible Shrinking Man* (1957), *Honey I Shrunk the*

---

[6] Which, as subtitles go, is right up there with Sharknado 2: The Second One.

Kids (1989), and many others[7]. In the 2010 *Eureka* Christmas Special, entitled "O Little Town", it was the entire town of Eureka that shrank[8]. Apparently their motto is "Go small or go home."

Fantastic Voyage is probably the most influential, in part due to taking a popular audience inside the workings of the human body for the first time[9] as the film follows a miniaturized medical team trying to save the life of a patient, and in part due to Raquel Welch performing her breakout role as one of the medics.

The problem with miniaturization is that there are two ways in which this might be accomplished, both outlandish in their own ways. The first involves interatomic separation. As we've discussed, the state of all subatomic particles can be described by a set of *quantum numbers*. To characterize an electron orbiting an atom requires four quantum numbers. One describes the energy level of the electron—its orbit. Two numbers describe its angular momentum, and the final number describes the "spin" of the electron. A rule called the *Pauli exclusion principle*, states that two electrons within the same atom cannot have identical sets of quantum numbers. It is for this reason that atomic orbits do not overlap, and most of the matter in our daily experience is empty space[10]. The quantum nature of electrons also means that you can't make the orbits themselves smaller without some serious tinkering to fundamental physical constants—sufficiently so that the shrunken atoms would stop behaving like the kind of particles that are required to make a functioning spaceship, or even a functioning Raquel Welch.

Conceptually, compressing the distance between atoms is one way that miniaturization might be achieved. In *Star Trek: The Animated Series* (1973–1974) there was an episode entitled "The Terratin Incident," in which a beam of unknown origin strikes the *Enterprise*, causing the crew to shrink. Spock offers the explanation that it was the gaps between organic molecules—the crew's bodies and their algae-based xenylon uniforms—were decreasing.

Reducing interatomic distances conserves mass. Alternately, an object might shrink if its mass was decimated, simply removed from the shrinking object. Neither of these prospects is scientifically satisfying. Science fiction novelist, and UC Irvine physicist, Dr. Gregory Benford summarizes, "I don't understand remotely how [miniaturization] could work. You can't do it by conserving the mass, because the obvious thing is that the *Incredible Shrinking Man* would have such a high density that he'd fall right through the floor. So you can't conserve mass when you're shrinking, you have to take stuff out and, in that case, if you shrink a human being down to 100th or 1000th of his size, then … what's he going to think with? He's missing over 99 % of his neurons. He's not even a human being anymore. So none of this makes sense to me."

Natural miniaturization does occur, in a sense, in dying stars. In larger stars nearing the end of their lifetimes, the crush of gravity overcomes the Pauli exclusion principle. Atoms are squeezed under such intense pressure that electrons and protons are literally forced together to form neutrons. Even though neutrons have no charge and do not experience electrostatic repulsion, further collapse is prevented

---

[7] In particular the holosuite, which would have to have colossal amounts of memory to perform as depicted.

[8] In fact, in *Stargate SG-1* the character of Samantha Carter made explicit distinction between a quantum singularity, with the implication that these are artificially constructed, and a "natural black hole."

[9] While the authors were writing the first draft of this book, *Doctor Who* aired "Into the Dalek." Ouch. The antibodies were pretty cool, though.

[10] Given, though, that the episode was told in flashback form by Sheriff Carter, one has to wonder if there ever really was any actual shrinkage.

by the *degeneracy pressure* of the neutrons, created again by the Pauli principle. If the gravity is so strong as to overcome even this pressure, nothing can stop total collapse. Neutrons can be packed one up against the other, creating a material so dense that a spoonful in Earth's gravity would weigh as much as a mountain.

It is only when one physical phenomenon trumps another that miniaturization can occur. One might argue that a narrative populated with nothing but neutrons might be dramatically uninteresting.

Let's say we want to teleport an atom, which we'll call A. Then we create a pair of entangled particles. The entangled pair is divided between two chambers—call them chamber X and chamber Y. Then atom A is put inside chamber X, where it enters into a quantum superposition with an entangled particle. In effect, this imprints the entangled particle with information about A. Then a measurement is made of the superposition, which extracts some classical information about the combined state of atom A and entangled pair. This does two things: the act of making a measurement causes the entangled atom in chamber Y to take on properties corresponding to its pair in chamber X, and it completely scrambles the quantum numbers of A.

This last effect is due to a feature of quantum mechanics known as the *no cloning theorem*. This states that it's impossible to determine an object's quantum state in any way that would allow two or more copies to exist simultaneously. Sadly, this puts the kibosh on any number of transporter accident-based screenplays in which duplicates (evil or otherwise) are created, such as the 1993 "Second Chances" episode of *Star Trek: The Next Generation*, where Commander Riker discovers a version of himself that was stranded on a planet during an evacuation several years previously.

The measurement information is then transmitted to chamber Y using any regular communications system, such as radio. Then a "blank" atom is fed into that chamber, where it interacts with the entangled particle already there. The measurement information transmitted by radio is then combined with the quantum information that came via the entangled particle, and the upshot is to transfer a complete set of quantum numbers, transforming the blank atom into a duplicate of atom A (Fig. 7.2). Simple, huh?

In the real world, the next step will be to find a way to teleport an entire molecule. Clearly, we're a long way away from teleporting a human, but perhaps one day this system could be used as a method of interstellar travel that would bypass the need to carry astronauts on a space ship that travels between the stars. Pairs of entangled particle pods could be made. One set of pods would be kept on Earth, while their pairs would be sent to another world along with a radio receiver, taking decades or centuries to cross the light

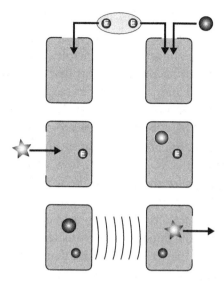

Fig. 7.2 Quantum teleportation of atoms requires two chambers that have been pre-prepared with a pair of entangled particles. To one chamber a "blank" atom is added. Then the atom to be teleported is introduced to the other chamber. It combines with the entangled particles there, scrambling its unique characteristics, and the result is measured and transmitted by conventional methods to the chamber with the blank atom. This copies the characteristics of the original particle to the blank atom

years between the stars. When the pods arrive at their destination, astronauts could step into the Earth pods, converting their bodies to an atomic mulch. The classical information about their bodies is beamed by radio to the planet, reaching it at the speed of light. The receiver combines the classical information with the particles in the landed pods. Assuming nothing goes wrong, and no entangled particles prematurely decohered during their journey, the astronauts would step out of the pods as if but an instant had passed since they were on Earth. This form of teleportation would share an ethical consideration with the teleportation method depicted in *Star Trek*: does it transport both the matter and essence of a living sentient being, or is it a murdering clone maker?.

Closer to reality, this technique is currently being studied as a way to absolutely securely transmit information between two points, albeit this has only been tested in the lab over a distance of a few meters. (This doesn't violate the rule against faster-than-light communication mentioned earlier, because classical information still needs to be sent at light speed or slower to make the system work).

Being able to teleport someone *without* the aid of entangled particles at the origin and destination would require scanning them at a level of accuracy that would violate Heisenberg's uncertainty principle. No-one much cared about this detail of physics when teleportation was first introduced to *Star Trek* in the 1960s as a time- and money-saving device. Fans with a knowledge of physics had 30 years to point out issues with the science of the original *Star Trek*, so in the 1990s "Heisenberg compensators" started appearing in the script technobabble regarding *Star Trek*'s transporters.

While the show provides no details on how Heisenberg Compensators function, it's a good demonstration of how the spread of scientific knowledge among fan bases can have an impact on how a show is written.

The more recent *Star Trek* series have also acknowledged that teleportation is primarily an information game. When characters are transported, they are temporarily stored in the "pattern buffer," and that information degrades rapidly over time, in much the same way that it is difficult to maintain a superposition within a quantum computer. (Although there are exceptions to this degradation effect, such as in the 1992 "Relics" episode of *Star Trek: The Next Generation*, (Fig. 7.3) in which Scotty is recovered from a modified pattern buffer after 75 years in storage.) In the 1995 "Our Man Bashir" episode of *Star Trek: Deep Space Nine*, a transporter malfunction involving several crew members requires saving the contents of the pattern buffer in the memory of every available subsystem within Deep Space Nine's main computer.

Science Box: Storing the Information to Teleport a Person

Teleporters, transportation staples of science fiction series like *Star Trek*, *Blake's 7*, and *Doctor Who*, are said to convert matter (to wit: a person) into pure energy, "beam" that energy to a destination, then reconstruct the original matter. It might be best if nobody got their hopes up for this to solve anybody's commuting woes, as the concept of teleporters, as typically described in works of science fiction, are fraught with technical problems.

We have already addressed the issues with the Heisenberg uncertainty principle, but there are others that are equally problematic. Let's say that a person's mass is 100 kg (true for both authors), that mass, converted to energy, would be

$$E = mc^2$$

or

$$E = (100\,\text{kg})\left(3\times10^8\,\text{m}/\text{s}\right)^2,$$

$$E = 9\times10^{18}\,\text{Joules}.$$

Assume that it is possible to construct a device capable of storing roughly 40 times the energy released by the largest nuclear explosion in history. Assume that there is a workaround so that the system does not violate the Heisenberg principle, there is still a data storage issue. The system would have to store enough information to recreate a person. How much information is that? If we had a huge cache of terabyte

Fig. 7.3 Engineers Georgi LaForge and Montgomery Scott on the Transporter pad from the *Star Trek: The Next Generation* episode "Relics" Copyright © Paramount Television. Image credit: Photofest

drives (reasonably large at the time of this writing, so yesterday's news in a year or two), how many would it take?

In order to get around the issue with the Heisenberg principle, let's assume that our transporter has a subroutine in its firmware where the input is an element, and a set of coordinates to build one atom of that element, without having to place exactly the subatomic particles that compose it. So for every atom in your body you would need a data structure like:

Atom type	Integer (2 bytes)
X coordinate	Double precision (8 bytes)
Y coordinate	Double precision (8 bytes)
Z coordinate	Double precision (8 bytes)

So we need 26 bytes to characterize the element and position of every atom in your body.

The top 13 elements, over 99.9 % of a person's mass, would be partitioned accordingly (recall that this is for a 100 kg person):

Element	Mass (kg)	Moles of element	Number of atoms	Bytes of storage
Oxygen	61.43	3839.61	2.31E+27	6.01E+28
Carbon	22.86	1903.26	1.15E+27	2.98E+28
Hydrogen	10.0	9921.62	5.98E+27	1.55E+29
Nitrogen	2.57	183.48	1.11E+26	2.87E+27
Calcium	1.43	35.68	2.15E+25	5.59E+26
Phosphorus	1.11	35.84	2.16E+25	5.61E+26
Potassium	0.20	5.12	3.08E+24	8.01E+25
Sulfur	0.20	6.24	3.76E+24	9.77E+25
Sodium	0.14	6.09	3.67E+24	9.54E+25
Chlorine	0.14	3.95	2.38E+24	6.18E+25
Magnesium	0.03	1.23	7.43E+23	1.93E+25
Iron	0.006	0.11	6.47E+22	1.68E+24
TOTAL	–	15942.23	9.60E+27	2.50E+29
Tb	–	–	–	2.27E+17

So it would take roughly 2.27×10^{17} terabytes to store the data to transport a 100 kg person.

Say a 1 Tb drive is $2 \times 8 \times 12$ cm (192 cm^3 or 1.92×10^{-4} m^3). Then 2.27×10^{17} terabyte drives would take up a volume of 4.36×10^{13} m^3. The volume of the Empire State Building is approximately 1 million (10^6) cubic meters. Therefore, by today's definition of "typical" for a 1 Tb drive, to accommodate the data necessary to transport one 100 kg person, it would require storage that would fill 4.36×10^7, or 43.6 million, Empire State Buildings. "Big data" may be all the rage these days, but not *that* big.

7.6 Quantum Creations

Ironically, as science fiction slowly gets better at paying attention to the ins and outs of real quantum mechanics, and exploiting them for storylines, purportedly non-fiction examinations of the physics have displayed a tendency towards pseudoscience. Part of the problem is in the nature of quantum mechanics itself. Many features of quantum mechanics, such as superposition and the uncertainty principle, don't correspond to anything that is familiar in everyday reality. Consequently, much of quantum mechanics is conceptualized by making analogies, even by scientists. Take the electron's quantum property of "spin" mentioned earlier. Quantum spin got its name because the equations that describe it are similar to those used to describe a property of macroscopic objects called spin, or intrinsic angular momentum—a property of a spinning planet or merry-go-round. Quantum mechanical spin is a fundamentally unimaginable property: unlike classical angular momentum, which can take on any value (well, up until the point that the object is spinning so fast that centrifugal force tears it apart), an electron's spin can only take on one of two values: $+ 1/2$ or $- 1/2$ (sometimes referred to as "up" and "down" spin). There is no such thing as an electron with *no* quantum spin, or one with 1/3 spin.

The danger arises when these analogies are taken to have literal meaning in the macroscopic world. For example, in 1979 Gary Zukov published a book called *Dancing Wu Li Masters*, which is actually a solid non-mathematical introduction to quantum mechanics. Throughout the book however, principles from Eastern mysticism are used to illustrate quantum mechanical concepts, and sometimes the line between analogy and physics gets blurred. Quantum mechanics has become a recurring theme in certain New Age philosophies ever since.

In this way, quantum mechanics is often invoked to give a veneer of plausibility to debunked ideas such as telepathy or astral projection. In the 2004 "documentary" *What the #$*! Do We Know!?*, computer graphics and "expert" interviews were comingled with a fictional narrative that suggests that human consciousness is so deeply intertwined with quantum mechanics that we influence reality by thinking about it hard enough. (The film is proof that, like electrons, movies can have quantum mechanical spin.)

In reality, there is a fundamental law of quantum mechanics called the *correspondence principle*, which states that in large systems that have a quantum mechanical description—systems of many particles or high energies—the quantum behavior must reproduce the classical physics behavior. Rephrased, quantum mechanical effects do not scale, and we have no fear that the real world may start behaving like *The Matrix*, where just thinking the right way about a spoon is enough to bend it like taffy.

In a similar fashion, the 2008 movie *Yesterday Was a Lie* features a neo-noir detective who finds herself in a surreal world, as multiple timelines begin to overlap (Fig. 7.4). James Kerwin, director of *Yesterday Was a Lie*, says, "While, obviously, we stretched the science a little bit, I thought that some of the concepts in quantum mechanics had similarities to the way that people view their personal relationships, particularly in their romantic life—the choices that we make and how the choices we make haunt us and can take us in different directions. So I used concepts of quantum mechanics as a metaphor for human relationships. So that's kind of the theme of the film: the nature of memory, the nature of time, the nature of reality, and the nature love."

There is a vast difference between using quantum mechanics as a narrative metaphor, as Kerwin does, and claiming that quantum effects manifest on the macroscopic level, as some pseudoscientists and pseudoscientific filmmakers do. Even the most willing among us to accept less-than-accurate science might find one "Oh, wow!" and the other "Oh, please".

Fig. 7.4 Chase Masterson (L) and Kipleigh Brown ® in *Yesterday Was a Lie* from Entertainment One. Photo by Josh Blakeslee. Copyright © 2009 Helicon Arts Cooperative. All rights reserved

> With *Yesterday Was a Lie* I was just trying to tell a good story. I let the science serve the story, not the other way around.
>
> James Kerwin, Director, *Yesterday Was a Lie*

Quantum mechanics is weird, but it is still science—our attempt to understand the fundamental underpinnings of our Universe. The randomness at its heart does not mean the Universe is a free for all. Just like playing roulette or poker in a casino, there are very definite rules that govern how it all plays out. God may indeed enjoy a good game of dice, but he is also a very strict pit boss.

Science Box: Seeing Quantum Singularities in a Hole New Light

One trapped the starship *Voyager* within its event horizon, others power Romulan Warbirds, still others power sharship-sized "supergates" in the *Stargate* franchise. *Quantum singularity* is a term that crops up fairly often in Hollywood science fiction and, like Schrödinger's wave function, the meaning of the term is not well defined across different Hollywood productions. Its use in context, however, allows viewers to collapse that wave function into a single definition—which is almost always equivalent in meaning to *black hole*[11]. Technical terms just seem so much more exotic and cooler when you prepend the term "quantum."

Just as there is a lot of uncertainty in Hollywood productions about what a quantum singularity might be, there is also a lot of uncertainty about what a black hole

---

[11] Kitty Pryde of *X-Men*, had the ability to travel through solid objects like walls by passing her atoms through the interstitial space of the atoms of the object through which she is moving. Her actual superpower, then, is that she can nullify or overcome the Pauli exclusion principle.

is, and what it can, and can not, do. Black holes are the supercompressed cores of dead stars, created when the crushing force of gravity overwhelms the quantum mechanical Pauli exclusion principle, and the matter that created the gravity collapses into a singularity.

While the details can get complex, a star is in essence a rather simple thing. It is a sphere of plasma and hot gas that generates its own energy through the process of nuclear fusion. A star lives its life in a state called hydrostatic equilibrium—the huge mass and gravitation of a star tends to compress its matter, but the incredible energy released from the fusion reactions in its core makes the matter tend to expand. A star's final size is determined by the outcome of this tug of war between gravitational collapse and an explosion.

As stars near the end of their lives (more on this in Chapter Eight: "My God, it's Full of Stars"), the rate of fusion slows, gravity wins the tug of war, and the only thing sustaining the star against gravitational collapse is the *degeneracy*[12] pressure generated by the Pauli exclusion principle as atoms and their electrons are squeezed together into one super-sized quantum system. Over time, as the energy output dwindles further, gravity trumps the Pauli principle, and the result is either a neutron star or a black hole.

A black hole was the "star" in Disney's 1979 film *The Black Hole*. The film was beautifully shot, and ground-breaking on several movie-making fronts, but obsessive attention to scientific detail was not one of them—even given the state of knowledge at the time[13]. The film is a product of its time, and the type of harsh science critique a film might expect today might be considered unfair since, as writer/producer Rockne O'Bannon reminded us in Chapter One, back then science simply had to pass the "smell test": if it seemed real, that was good enough. Nevertheless, *The Black Hole* repeatedly propagated the most common misconceptions about black holes: that they are terrifying insatiable cosmic kaiju that devour everything in their surroundings. We can use the film to explore the realities of black holes.

Early in the film, upon first seeing the eponymous black hole, Dr. Alex Durant (Anthony Perkins) describes it as, "The most destructive force in the Universe." Later, mad scientist Dr. Hans Reinhardt (Maximilian Schell) asks Durant, "Are you interested in black holes?" Durant replies, "How can one not be overwhelmed by the deadliest force in the Universe?" Some stellar explosions, are so large, and release so many gamma rays, that they literally blow apart their own black hole in the process of formation. This clearly calls into question whether a black hole really represents the most destructive force in the Universe. The Universe can be a scary place, and there are many ways in which the Universe can conspire to end us that are far more effective than black holes[14](Fig. 7A.1).

Regarding the black hole, Durant claims, "Nothing can escape it, not even light," and, shortly after, "...you can see giant suns sucked in and disappear without a trace." It is true that, close enough to a black hole, the gravitational attraction is so great that even light can not escape, and astronomers have witnessed the apparent digestion of stars by black holes. A black hole swallowing a star is, however, a

---

[12] Two subatomic particles, like electrons, are said to be degenerate when they have the exact same quantum state (which is forbidden for many such particles). Try to force them into the same quantum state, they push back.

[13] When the Earth spaceship *Palomino* comes across the (apparent) derelict *Cygnus*, Dr. Kate McCrae (Yvette Mimieux) says of *Cygnus*, "Its mission, to discover habitable life in outer space. Same as ours." Habitable life? Taken literally, this statement makes humans out to be space parasites.

[14] If you weren't planning on sleeping anyway, try our friend Phil Plait's book, *Death from the Skies: These Are the Ways the World Will End*.

Fig. 7A.1 The crews of *Cygnus* and *Palomino* observe a black hole in Disney's *The Black Hole* (1979). Copyright © Walt Disney Productions. Image credit: Photofest

rare and violent act that's accompanied by considerable astronomical pyrotechnics. Matter may not be visible after it enters a black hole, but it can make one heck of an X-ray light show on the way down. Also black holes do emit a drizzle of particles in the form of Hawking radiation due to the Heisenberg uncertainty principle. This drizzle causes the radius of the black hole to shrink. It's a minute effect for a black hole with a stellar mass or more. Nonetheless, in the unimaginably distant future, it means that even the supermassive black hole at the heart of our galaxy that we discuss below will simply evaporate into nothing, all due to a tiny quantum effect.

> I think that a really important thing is that probably 90 % of the time you don't need to choose between good science and good story. If anything, if you care deeply enough about the science, and care deeply enough about the premise you're using, you actually find really cool story points that pop out of the physics and the chemistry and the laws of the universe.
>
> Zack Stentz, Writer/Producer

In the *Doctor Who* Episodes "The Impossible Planet" the Doctor is gobsmacked by the existence of a planet in orbit about a black hole, and repeatedly emphasizes that the planet should not exist. In reality, there is nothing in the laws of gravity or orbit dynamics preventing it as viewers saw in *Interstellar* (2014). Recall that a black hole has the mass of the core of its progenitor star, not even the entire star. At the distance, planets would orbit the star during its main sequence lifetime, a black hole would have significantly less gravitational attraction. The primary problem for such a planet or, more to the point, any base, habitat, or installation on such a planet, would be that it would get pummeled by material falling into the black hole—which was exactly what was happening in "The Impossible Planet" (and the second of the two-part episode, "The Satan Pit"[15]), but which was a life-threatening complexity not depicted in *Interstellar*.

---

[15] In "The Impossible Planet"/"The Satan Pit" episodes of *Doctor Who*, and the movies *The Black Hole* and *Event Horizon* (1997), black holes are associated with Hell or Satan himself. In fact, a better metaphor for black holes is that they are cosmic roach motels: you check in, but you don't check out.

What makes a black hole dangerous is that the matter is packed so much more tightly than is possible for a main sequence star (see Chapter Eight: "My God, It's Full of Stars: The Universe"), that objects can approach far closer and not get fried[16] by starlight or blown away by stellar winds. So objects can get close enough to the mass to become trapped forever. The event horizon of a black hole is a sphere at the exact distance where the escape velocity equals the speed of light—and nothing with mass can travel the speed of light. How far away from the black hole is the event horizon? We start with the equation for escape velocity, the minimum velocity required to escape any object with mass:

$$v_e = \sqrt{\frac{2GM}{r}},$$

where v_e is the escape velocity, M is the mass, r is the radial distance from the center of the object, and G is the gravitational constant we saw back in Chapter Four: "Matter Matters." If the event horizon is where v_e is equal to the speed of light c, or 3.0×10^8 m/s, then this becomes:

$$c = \sqrt{\frac{2GM}{r}}.$$

The radius of the event horizon of a black hole is known as the Schwarzschild radius, or R_s. Replacing r with R_s, rearranging, and solving for R_s:

$$c^2 = \frac{2GM}{R_s},$$

or,

$$R_s = \frac{2GM}{c^2}.$$

Now that we have an equation, what is the Schwarzschild radius for a 10 solar mass black hole? The mass of Sol is 2×10^{30} kg, so for a 10 M_{sun} black hole, the mass is 2×10^{31} kg, and:

$$R_s = \frac{2(6.67 \times 10^{-11} \frac{m^3}{s^2 \cdot kg})(2.0 \times 10^{31} kg)}{9.0 \times 10^{16} \frac{m^2}{s^2}} = 2.96 \times 10^4 m.$$

Just under 30,000 km! This is only 4.6 times the radius of Earth. Communications and weather satellites in Earth geostationary orbit are outside of this distance at 35,800 km.

In *The Black Hole*, Dr. MacCrae relates, "I had a professor who predicted that, one day, black holes would devour the entire Universe." Perhaps, but the Universe might end before that can happen. Astronomers believe that, at the core of the Milky Way, there is a supermassive black hole, whose estimated mass is 4.3 million solar masses, or 8.6×1036 kg. The radius of its event horizon would be

$$R_s = \frac{2(6.67 \times 10^{-11} \frac{m^3}{s^2 \cdot kg})(8.6 \times 10^{36} kg)}{9.0 \times 10^{16} \frac{m^2}{s^2}} = 1.27 \times 10^{10} m = 1.27 \times 10^7 km$$

---

[16] Well, actually, nearby objects would get fried, but by the radiation released by infalling matter, not by the energy release of a main-sequence star.

That may seem like a large number, but if the largest black hole in our Galaxy replaced the Sun, the event horizon would extend only about 20 % of the way to the planet Mercury.

The king of cinematic black holes would be Gargantua, one of the… stars… of *Interstellar*. Gargantua was depicted as a 100 million solar mass ($2.0 \times 10{38}$ kg) black hole[17]. The radius of its event horizon, then, would be

$$R_s = \frac{2(6.67 \times 10^{-11} \frac{m^3}{s^2 \cdot kg}) \, (2.0 \times 10^{38} \, kg)}{9.0 \times 10^{16} \frac{m^2}{s^2}} = 2.96 \times 10^{11} m = 2.96 \times 10^8 km,$$

or about twice the distance between the Sun and Earth.

Like so many other hazards in Nature, a black hole is dangerous only if you get too close—which means that "too close" is where the drama lies (See also the science box "The Physics of Miller's Planet" in Chapter Eleven). In "Parallax", the second episode of *Star Trek: Voyager* (1995–2001), the crew of the starship *Voyager* find themselves within the event horizon of a quantum singularity which clearly, in context, is a black hole.

Clearly the writers of the episode knew that the event horizon of a star represented the "point of no return"—the point where not even light can escape the black hole—but didn't know *why* it was the point of no return. In what amounted to a major "Oh, please!" moment for many viewers, the crew of *Voyager* found a "crack" in the event horizon, and made their escape through that. The problem is that for a warp- or FTL-capable (faster than light, see Chapter Nine: "Shortcuts through Time and Space") craft like *Voyager*, the event horizon is meaningless. *Voyager* had merely to accelerate to warp speed, and this should have been the shortest episode of *Star Trek* ever[18].

These two examples underscore an important point. Science is about explaining the "whys" and the "hows" of Nature and rarely just about the "whats". Knowing a scientific fact, and knowing why it is a fact, and how science came to understand it as a fact, can often be the difference between an "Oh, wow!' and an "Oh, please!" moment.

---

[17] Thorne K. (2014) The Science of Interstellar. W. W. Norton and Company, Inc.
[18] That is if Voyager's duranium hull had the tensile strength to keep her from being torn apart at the atomic level by tidal force due to its proximity to the black hole.

8

My God, It's Full of Stars: The Universe

When we try to pick out anything by itself, we find it hitched to every-thing else in the universe.
John Muir, *Naturalist*

We believe that the universe itself is conscious in a way that we can never truly understand. It is engaged in a search for meaning. So it breaks itself apart, investing its own consciousness in every form of life. We are the universe trying to understand itself.
Minbari Ambassador Delenn, *Babylon 5*, "Passing through Gethsemane"

All you really need to know for the moment is that the universe is a lot more complicated than you might think, even if you start from a position of thinking it's pretty damn complicated in the first place.
Douglas Adams, *Mostly Harmless*

In the original 1978 *Battlestar Galactica*, after the Cylon onslaught lays waste to the Twelve Colonies of Humanity[1], the battlestar *Galactica* shepherds a rag tag fleet of over 200 vessels[2] of all shapes and sizes away from their ravaged homeworlds, and across space in search of a new home. Commander Adama tells the Council of the Twelve that all is not lost, and he has a plan, saying:

---

[1] It's worth watching for the first 40 minutes alone, the attack on the Colonies holds up really well.
[2] In the 2003 reimagined version, it was closer to 60.

> ADAMA
> Where will we go? Our recorded
> history tells us that we descended
> from a mother civilization: a race
> that went out into space to
> establish colonies. Those of us
> here assembled now represent the
> only known surviving colonies…
> Save one. A sister world far out
> in the universe, remembered to us
> only by ancient writings. It is my
> intention to seek out that
> remaining colony, that last
> outpost of humanity in the
> universe…. I wish I could tell
> you that I know precisely where it
> is, but I can't. However I do know
> that it lies beyond our star
> system, in a galaxy very much like
> our own, on a planet called…
> Earth[3].

Space isn't remote at all. It's only an hour's drive away if your car could go straight upwards.

"Sayings of the Week", *The Observer* (9 September 1979)

Over the course of the 21 episodes of the original series, it is established that flank speed for a battlestar is light speed. In the reimagined version of the show, *Galactica* is even faster, capable of near-instantaneous jumps of up to five light years, and yet over the span of 78 episodes, travels only

---

[3] A recurring theme of the reimagined *Battlestar Galactica* was, "All of this has happened before, and all of this will happen again." Clearly, in the most recent telling of this recurring drama, Commander Adama not only had a much better grasp of the nature of the Universe, but he was also a much better motivational speaker. So Say We All

1/10th of the way across this Galaxy, tops.[4] If Earth in the original series was indeed to be found "far out in the universe… in a galaxy much like our own," then nobody aboard would survive the trip because it would take millions to billions of years[5]. Clearly, the writers of the original series were unaware of the difference between a star system, a galaxy, and the Universe, because more than once, the terms are used synonymously. In order to understand those distances, and why there is so much space in, well, space, we need to go right back to the beginning, where everything starts with a bang.

Science Box: The Size of a Galaxy vs. a Star System

A star system is simply a star, or a few stars bound together by gravity, along with the objects that are orbiting them, like brown dwarfs, planets, asteroids, and comets. A star system may span dozens of *astronomical units*, or AU, across. An astronomical unit is the measure of distance used within the Solar System, and a simple definition is that it is the average distance from Earth to the Sun[6]. Using this measure, distances range from reasonable fractions of an AU (from the Sun to Mercury is roughly 0.4 AU), to tens of AU (from the Sun to Neptune is about 30 AU). The band of icy bodies beyond Neptune, the Kuiper Belt, extends to almost 50 AU.

We can approximate the size of the Solar System as a disk 50 AU in radius and 1 AU thick. The volume of a disk is:

$$V = \pi r^2 h,$$

---

[4] The math behind this is worked out in detail in DiJusto and Grazier, *The Science of Battlestar Galactica*, Wiley, 2009.

[5] Given that the Universe is ever-expanding, if it was far enough out into the Universe, Earth could be flat-out unreachable.

[6] The definition is actually much more complex (isn't it always?), but this definition is close.

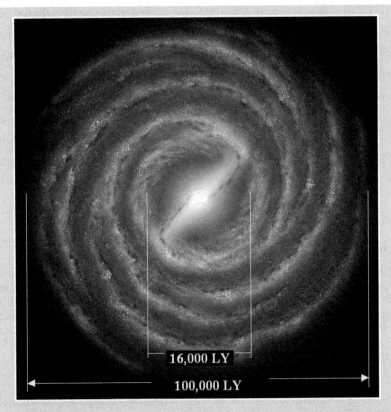

16,000 LY

100,000 LY

Fig. 8A.1 An artist's conception of the Milky Way. (Photo Credit: NASA/JPL-Caltech)

where *r* is the radius, and *h* is the height. So the volume of the Solar System is:

$$V = \pi \left(50 \text{ AU}\right)^2 \left(1 \text{ AU}\right),$$

Which is 7854 cubic Astronomical Units, or 1.74×10^{20} cubic kilometers.

A galaxy, on the other hand, may have millions, billions, even trillions of star systems, and is thousands to hundreds of thousands of *light years* across. The term "light year" can be confusing because, despite the name, it not a measure of time, but rather a measure of distance—one light year is the distance a beam of light can travel in one year. Given the speed of light, and the fact that a year is 31,557,600[7] seconds long, simple multiplication shows that one light year is equal to a little under 9.5 trillion kilometers, or 5.8 trillion miles.

Figure 8A.1 shows an artists conception of the Milky Way Galaxy. It is approximately 100,000 light years across. In other words, the same beam of light that travels around Earth's equator 7 times in a second, travels from Earth to the Sun in 8 ½ minutes, would take 100,000 years to cross our Galaxy.

---

[7] An approximation, and mnemonic, that physicists use is that one year is $\pi \times 10^7$ seconds.

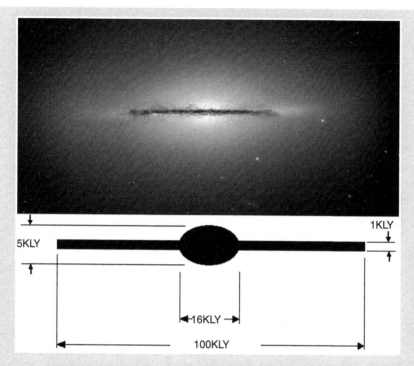

Fig. 8A.2   NGC 5866, a galaxy shaped like ours seen edge-on, and an idealized representation showing the dimensions of the Milky Way. (Photo Credit: NASA, ESA, and Hubble Heritage (STScI/AURA))

If a crew were to travel through space at the speed of Apollo 10—which became the fastest crewed spacecraft to date when it re-entered Earth's atmosphere at a hair over 11 kilometers per second— it would take them 2.7 billion years to cross the Milky Way. Our closest galactic neighbor—M31 or the Andromeda galaxy—is about 23.5 billion billion kilometers away. Apollo 10 would take *67.7 billion years* to reach Andromeda. By comparison, recall that the universe is only 13.8 billion years old.

Seen edge-on, our Galaxy has a central bulge, 16,000 light years by 5,000 light years, but the disk region containing the spiral arms is, comparatively speaking, extremely thin—only 1000 light years thick. The Solar System is in a spiral arm about 60 percent of the way from the galactic core to its periphery. Figure 8A.2 shows a galaxy like ours edge-on. In this idealized model we have an oblate spheroid, i.e., the bulge and bar, immersed within a disk. The volume of an oblate spheroid is:

$$V = \frac{4}{3}\pi a^2 b$$

Where *a* is the semi-major axis of the spheroid, and *b* is the semi-minor. In this case of the galactic core, *a* is 8000 LY, and *b* is 2500 LY. The volume of the core is:

$$V_c = \frac{4}{3}\pi \left(8000 \text{ LY}\right)^2 \left(2500 \text{ LY}\right),$$

or,

$$V_c = 6.70 \times 10^{11} \; LY^3.$$

So the volume of the galactic core and bar is 6.7×10^{11} cubic light years. The volume of a disk is:

$$V = \pi r^2 h$$

Where r is the radius, and h is the height. In this case, though, we have to subtract the volume from the center of the disk, which has already been accounted for by the galactic core. So the adjusted volume is:

$$V = \pi h \left(r_{galaxy}^2 - r_{core}^2 \right)$$

or

$$V_d = \pi (1000 LY)(2.5 \times 10^9 \, LY^2 - 6.4 \times 10^7 \, LY^2),$$

or,

$$V_d = 7.65 \times 10^{12} \, LY^3.$$

The total volume, then, is simply the sum of the bulge and disk volumes

$$V_{MW} = V_c + V_d = 8.32 \times 10^{12} \, LY^3.$$

Or just over 8 trillion cubic light years, or 7.0×10^{51} cubic kilometers. When Commander Adama, or the writer who speaks through him, mistakes a star system for a galaxy, he is off by a several hundreds of billions of stars, and a factor of 40 thousand billion billion billion in volume.

8.1 An Even Briefer History of Time

Sir Fred Hoyle[8] was the Carl Sagan of his day. He was an astronomer, he wrote science fiction, and he was a character. He also did not believe that the Universe had a beginning, but rather the Universe had always existed and always would exist. In one of the most noteworthy science/media interactions ever, when an interviewer on a BBC radio broadcast in March 1949 raised the increasingly popular notion that the Universe has a beginning, Hoyle bristled, "These theories were based on the hypothesis that all the matter in the uni-

[8] Fred Hoyle was something of an enigma. Some of his views would be considered modern today, yet some of his views bordered on intelligent design. http://www.joh.cam.ac.uk/library/special_collections/hoyle/exhibition/radio/

verse was created in one big bang at a particular time in the remote past."
Though originally a pejorative term[9], scientists who embraced the theoretical
work of Georges Lemaître and the empirical observations of Edwin Hubble
enthusiastically adopted Hoyle's term "Big Bang" for the event that created
our Universe.

Hoyle found the idea that the universe had a beginning to be pseudoscience, despite mounting theoretical and observational supporting evidence
throughout the years, including the discovery of the *cosmic microwave background*, which is the afterglow from the Big Bang itself, in 1964. In short, the
whole notion sounded a little too much like "Let there be light!" for the tastes
of Hoyle and some other astronomers. He said, "…it's an irrational process,
and can't be described in scientific terms." Fred Hoyle believed in the steady-
state model of the Universe until his death in 2001.

> Equipped with his five senses, man explores the Universe around him and calls
> the adventure Science.
>
> Edwin Hubble, Astronomer

The narrative of our understanding of the Big Bang actually starts much
earlier. We'll pick it up mid-way through the 18th century with French astronomer Charles Messier. Messier was a comet hunter, yet kept observing
fixed fuzzy objects in the night sky that could easily be mistaken for distant
comets. To help himself, and other comet hunters, from repeatedly misiden-
tifying these objects, Messier compiled an initial catalog of 45 objects, which
later grew to 110. These *Messier objects* turned out to be star clusters, nebulae
(clouds of gas and dust in space), and galaxies[10] (although galaxies where initially mistakenly classified as nebulae or star clusters).

In the early twentieth century, astronomers believed that the Milky Way
comprised the entirety of the universe, with Sol near its center. Astronomers
also generally thought that the stars were eternal, and that the Universe had
no beginning or ending. When astronomers later realized that the physics of
nuclear reactions dictated that stars must age and die, the steady-state theory

---

[9] Though Hoyle later insisted otherwise.
[10] Many of the Messier objects were studied by the 3rd Earl of Rosse, William Parsons, who first discovered the spiral form of M51 (meaning the 51st object in the Messier catalog) in 1845. M51 is better
known today as the Whirlpool Galaxy. Parsons used a 1.8-meter telescope that he had constructed on his
estate in the middle of Ireland—this was the largest telescope in the world until the 2.5 Hooker telescope
was built on Mount Wilson in 1917. Edwin Hubble would use the Hooker telescope to perform his
groundbreaking research on the nature of galaxies.

was developed by Hoyle and others, which stated that new matter was continuously coming into existence to replace the raw material consumed by stars.

When Albert Einstein published his General Theory of Relativity in 1916, in which gravity is not so much a force exchanged between bodies with mass (the Newtonian view), but rather a warping of the fabric of spacetime caused my the mass of objects, he created serious cognitive dissonance for astronomers and physicists. It also spelt the beginning of the end for the steady-state theory. General Relativity predicted that the steady-state Universe would collapse because objects would gravitationally attract one another into, ultimately, one large mass. Since that did not fit the prevailing astronomical understanding at the time, and in the absence of empirical observations to the contrary, Einstein contrived the *cosmological constant*, what scientists call a "fudge factor." The cosmological constant was a repulsive force Einstein added to his equations of General Relativity in order to force his theory to fit the steady-state model, but it wasn't too long before new observations came in that argued in favor of rejecting the steady-state model in favor of an expanding and evolving universe. This prompted Einstein to (purportedly) call the cosmological constant his "greatest blunder"[11,12].

Unravelling the true nature of many of the Messier objects was a key step towards the end of the steady-state model. As telescopes improved, some of the Messier objects were observed to have spiral structures, and were called "spiral nebulae," but all Messier objects were still assumed to lie within the bounds of the Milky Way. By the 1920s, astronomers were debating if these spiral nebulae were actually "island universes," that is, gravitationally-bound ensembles of millions, even billions, of stars, just like the Milky Way. Astronomers had measured the distance to some of the stars in the Milky Way as far back as the nineteenth century, but in 1924 Edwin Hubble single-handedly increased the size of the known Universe by a factor of thousands. Using the new 100-inch Hooker[13] telescope atop of Mount Wilson just north of Los Angeles[14], the largest in the world at the time, Hubble determined that the distance to M31 was at least ten times farther away than the

---

[11] Or, perhaps, he didn't. While he certainly regretted the inclusion of the cosmological constant in the equations for General Relativity, it is unlikely he actually ever said one of his most oft-cited quotes: http://www.theatlantic.com/technology/archive/2013/08/einstein-likely-never-said-one-of-his-most-oft-quoted-phrases/278508/

[12] Given more recent observations, the cosmological constant may have been more prescient than blunder.

[13] That would be the name of a benefactor, not a profession, in this case.

[14] Hubble was a science celebrity, and made good use of his time in Los Angeles to party with Hollywood movie stars like Charlie Chaplin and Greta Garbo, and well as famous writers like Aldous Huxley. Hubble's wife, Grace Burke-Lieb, was active in shaping his public image and social calendar. In today's vernacular, she was, essentially, his publicist.

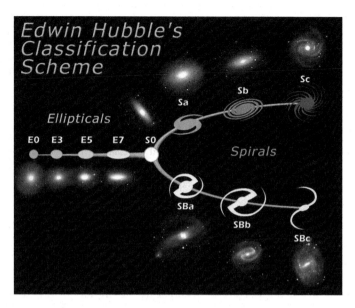

Fig. 8.1 Hubble's "Tuning Fork" diagram. (Image credit: STScI)

most distant stars of the Milky Way. Thus the Andromeda Nebula became the Andromeda Island Universe, now known as the Andromeda Galaxy[15]. Soon after, other astronomers determined that M31 was at least twice as far away as Hubble's estimate—our current best estimates put it at 2.5 million light years away, while the diameter of the Milky Way is about 110,000 light years.

In the intervening years, Hubble observed many galaxies, and in 1926 published a taxonomy of galaxies, one that has come to be called the "Tuning Fork Diagram" by astronomers (Fig. 8.1). Hubble classified galaxies into ellipticals, ranging from nearly spherical galaxies to some very elongate, spirals (a central bulge with arms radiating spirally outwards), and barred spirals (a central bulge with a bar passing through, and spiral arms attached to the end of the bar).

Hubble believed that his galaxy classification scheme depicted an evolutionary process—that galaxies started as elliptical and became spiral over time. In fact, it is often the reverse. On the rare occasions when large galaxies collide and merge, the result is typically an elliptical galaxy (somewhat more common are collisions between a large galaxy and a dwarf galaxy: when this happens the dwarf galaxy is completely ripped apart, and often consumed, by the larger galaxy). The Milky Way and the Andromeda Galaxy will undergo

---

[15] M31 is known as the Andromeda Galaxy, because we see it in the constellation Andromeda. It is the most distant object in the Universe that you can see with the unaided eye.

a collision and merge in approximately 4 billion years, and computer models strongly suggest that the result will be an elliptical galaxy. If that does happen, the formation of new stars, planets, and even new life, will dramatically slow, perhaps even come to a halt (for more details, see Chapter Eleven: "Braver Newer Worlds").

There is another type of galaxy, called an irregular galaxy, which is not on Hubble's diagram. Just as the name implies, these galaxies are oddly shaped, perhaps due to being disrupted by "recent" passage near a larger galaxy. The Large and Small Magellanic Clouds (aka the "Clouds of Magellan") look like clouds in the Southern Hemisphere sky, but they are actually small, irregular[16] galaxies that orbit our own.

Another irregular galaxy is the Pegasus galaxy, which is a distant companion of the Andromeda galaxy. Pegasus was the setting for *Stargate: Atlantis* (2004–2009), and it's a good example of how far scientific literacy in screen science fiction has improved since the days of the original *Battlestar Galactica*. Not only was the extreme distance to Pegasus acknowledged (some 3 million light years) on screen, it was actually the driver for much of the drama of the early seasons—creating a single wormhole to Pegasus from Earth required so much energy that it was a one-shot-deal, leaving the initial expeditionary force alone in a new galaxy without backup or new supplies.

Hubble made a second type of observation that he published in 1929. Hubble confirmed earlier work that nearly all galaxies are moving away from ours, and determined the linear relationship between the distance to a galaxy, and its rate of recession. In other words, galaxies farther away are moving away faster.

One of the best ways to visualize what's going on is to imagine raisins mixed into bread dough in a baking dish. The raisins represent galaxies and the bread dough represents space. Throw the dish with the raisin dough into the oven and turn on the heat. The dough will begin to expand at uniform rate throughout the tin. Now pick one of the raisins (it doesn't matter which) to be our Milky Way. Imagine when we turn on the oven that there is another raisin/galaxy one centimeter away, and another at the 2 cm mark, and so on each centimeter out to 5 cm. The dough starts to rise and expand, doubling in size over the course of a minute. So even though each raisin/galaxy remains at rest relative to the dough around it, the distance between it and the next raisin/galaxy expands from 1 to 2 cm. So we would say that the nearest raisin/galaxy to the Milky Way raisin is moving away from us at speed of 1 cm/minute. Now let's look at the next raisin/galaxy along. It's now 4 cm away. The one after that is 6 cm away, and the one that was originally 5 cm away is now

---

[16] Though both the SMC and LMC have bars, and the LMC displays some evidence of spiral structure.

10 cm distant, which we, in our Milky Way galaxy, would perceive as moving away from us at a speed of 5 cm/minute.

This has become known as *Hubble's Law*, though some argue that it should be called Lemaître's Law, since Georges Lemaître published similar findings two years earlier than Hubble, but did so with poorer data and in a lesser-known journal. Still, the implications were clear. If the arrow of time was reversed, and time made to run backwards in our universe, galaxies would all converge. The Universe had a beginning—a beginning that, in 1949, Fred Hoyle would call the Big Bang. Our current best estimate places the Big Bang at 13.8 billion years ago.

> In the beginning, the universe was created. This made a lot of people very angry, and has been widely regarded as a bad idea.
>
> Douglas Adams, The *Hitch-Hiker's Guide to the Galaxy*

So just how did the Universe get started? That's a very good question, and one theoretical physicists are struggling to answer. From the instant of creation until 10^{-43} seconds (that's a decimal point, 42 zeroes, and then a 1, also known as the *Planck time*) after the instant of creation, physics, as we understand it today, simply does not work. However, after that point, physicists have a surprisingly good grasp of the series of events.

When the universe was created in the Big Bang, what exploded was space itself. In other words, every point in the immense vastness of the cosmos that we see when we look out into the night sky was once compressed into something smaller than an atom (George Lemaître called the progenitor of the Big Bang the "primeval atom.").

From one Planck time after creation, and for the next few seconds, the Universe was a superheated milieu of energy and particles blinking into, and out of, existence. Imagine a mixture of alcohol and water heated to steam. Then cool the steam. First water condenses, amidst a steam of alcohol. Then alcohol condenses. Cool the mixture still further, and water crystallizes. Drop the temperature still more and, eventually, the alcohol will crystallize. That is a rough analogy of what happened during this period—different particles formed as the Universe cooled and underwent many such transitions. In this chaotic period it is believed by most physicists—though not all—the Universe underwent a period known as inflation, in which the expansion of the universe accelerated enormously, increasing its diameter from the size of proton to anywhere from a few centimeters to a few meters in less than 10^{-32} seconds.

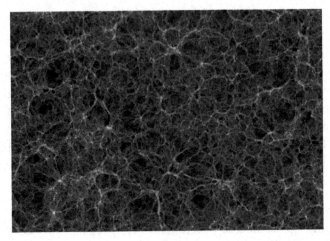

Fig. 8.2 Computer simulation revealing the large-scale structure of the Universe, showing the clumping imprinted from quantum fluctuations that grew under inflation. *Each dot* in the intricate spider web represents a galaxy. (Image credit: NASA)

For the next three minutes, the Universe was expanding and cooling, but at a slower rate, and particles like electrons, protons, and neutrons could form. At roughly the three minute mark, the temperature had cooled enough for the existence of atomic nuclei, and for the next 15 to 17 minutes, the infant Universe fused hydrogen to make helium nuclei and a tiny amount of lithium nuclei—though it was still too hot for the formation of atoms, which require electrons to be able to stay in bound orbits. It would take another 380,000 years for the Universe to cool enough for that.

The matter in the Universe was distributed fairly evenly over large scales, but "clumpy" over short scales. Microscopic quantum fluctuations (like the vacuum energy from Chapter Seven: "A Quantum of Weirdness") were magnified into macroscopic fluctuations during inflation. Any small "clump" of material in a region would have a gravitational edge over volumes with more uniform or rarefied mass densities, and would attract more material. This would create a virtual spiral as the more material drawn into a volume of space, the more mass it has, the more gravity is has, and the more material it draws.

Those clumps compressed still further until they formed galaxies' first stars about a 100 million years after the Big Bang (Fig. 8.2). The first stars were composed only of hydrogen and helium but, over time, the lives, and deaths, of these first stars created the heavier elements necessary for the formation of planets, organic compounds, and life.

When we observe galaxies today, we can see that space is still expanding from the Big Bang. Locally, gravity holds things together into planets, stars, and galaxies, but when we look out into deep space, as Edwin Hubble did in the 1920s, we see that the distances between galaxies are constantly increasing, as the galaxies are swept along by the expansion of the universe.

If we look at galaxies close to the edge of the observable universe, the measured separation between them and us is growing at a rate faster than the speed of light. One fact that Albert Einstein established in his Theory of Special Relativity, published in 1905, was that the universe had a hard speed limit, and no object could travel faster than light through a vacuum, which zips along at just under 300,000 kilometers per second[17]. The furthest object we have spotted so far is a galaxy 13.4 billion light years away, with an apparent motion away from us of nearly twelve times the speed of light. If you were to be magically transported to that galaxy it would appear to be stationary, and the Milky Way would appear to be the object moving at superluminal speed.

Odd as this seems, this does not violate the universal speed limit, because "speed" is a local phenomenon, and is defined by motion relative to nearby objects. Rappelling provides a good analogy. If you have ever rappelled down a rope of a significant length, you would have noticed that there was little spring in the rope at the beginning of the descent: rope isn't particularly stretchy and there is not much give. Say each foot of rope can stretch only 1/50th of an inch per foot. At the top of the descent, 10 feet down, there is only 1/5th of an inch "give" in the rope. By the time you are hanging from 100 feet of rope, though, the cumulative effect means there are 5 inches of spring in the rope, and that is certainly something you would notice.

Now imagine that, at the beginning of your descent, you notice two ants on the rope, just above your topmost hand—one is crawling down towards your hand, the other up and away. Both of these move away from their midpoint with velocity v, so their relative velocity to one another is $2v$. The ant crawling towards you climbs aboard your hand, and joins you in your descent down the rope.

After you have descended 100 feet, you gently lift the ant off your hand, place it on the rope so it is, again, crawling in the downwards direction. This time, though, you pull on the rope, and it stretches. Although, relative to the fibers in the rope, the ant at the top of the rope is still crawling at v, and the ant at the bottom is also crawling at v relative to the fibers at its end of the rope; as the fiber stretches, their relative velocity is $2v$ plus the rate of the rope stretch. Neither ant travelled faster than v relative to its local surroundings, but their relative velocity when they were distant exceeded their relative velocity when they were close.

What this implies is that, in the original *Battlestar Galactica*, if Earth was too "far out in the Universe," relative to the Twelve Colonies of Kobol, at

---

[17] Special Relativity also states that the Universe has no preferred frame of reference. Germane to the conversation here, it means that there is no point that is the "center" of the Universe. Phrased, differently, every point, and everybody, can lay equal claim to being the center of the Universe. For those who think "kids today" are too entitled, best not to relay this fact.

Fig. 8.3 False-color image of NGC 6744, a galaxy similar to the Milky Way, with a central bulge, a bar passing through that, and spiral arms radiating from the bar. (Image credit: European Southern Observatory)

"only" lightspeed, *Galactica* would not require millions or billions of years to reach it, it means that it could *never* reach it. Earth's Galaxy would be moving away far too quickly.

8.2 A Galaxy Not So Far Far Away: The Milky Way

As astronomers plotted the direction and distances of stars in our own Galaxy, a pattern started to emerge, arms were recognizable, and it appeared that the Milky Way was a spiral galaxy, Sb or Sc on Hubble's diagram. In the 1990s, as ground-based telescopes improved, and space telescopes like Hubble, Chandra, and Spitzer were launched, astronomers were able to see deeper into the Galaxy. They suspected that the Galaxy was actually a barred spiral. Observations from the Spitzer Space Telescope in 2005 confirmed that, rather than a simple spiral, our Galaxy is a barred spiral, more like an SBc from the Hubble classification scheme (Fig. 8.3).

The Milky Way is a larger-than-average Galaxy, roughly the same size as the Andromeda Galaxy, containing approximately 300 billion stars, and between 100 and 400 billion planets. Recent estimates suggest that the Milky Way may contain between 20 and 40 billion Earth-like planets, perhaps 10 billion of which orbit stars like Sol, our Sun.

Fig. 8.4 The vast scale of the Milky Way Galaxy becomes more apparent when we view it relative to the travels of the Starship *Voyager* and the Battlestar *Galactica*. If the *yellow dot* is the approximate position of the Solar System, *Voyager*'s 75,000 LY journey home is approximated by the *green arc*; all of *Galactica*'s travels occurred within the *red circle*. (Image credit: NASA/Caltech/JPL)

The Solar System[18] is about 3/5th of the way from the center of the Galaxy to its edge. The Solar System orbits about the center roughly every 240 million years. The last time Earth was in our current position relative to the rest of the Galaxy, the first dinosaur had yet to set foot on Earth—it was still the Age of Amphibians, and would be for another 15 million years.

To give a sense of scale to our Galaxy, we will look at the travels of two famous Hollywood spacecraft: the Federation starship U.S.S. *Voyager* and the (reimagined) battlestar *Galactica* (Fig. 8.4).

In the pilot episode of *Star Trek: Voyager* (1995–2001), the starship *Voyager* is thrown across the Galaxy, and the crew estimates it will take a trip of 75,000 light years—tracing an arc to avoid the radiation of the galactic core—and roughly 60 years to make it home. This is at high warp speed, several times the speed of light. The green arc in Fig. 8.4 shows, at least qualitatively, *Voyager*'s path home.

---

[18] Strictly speaking since there is only one star named Sol, there is precisely one Solar System (hence it is a proper noun and capitalized) in the Universe. Astronomers tend to call other planetary systems solar systems (a general term, hence not capitalized), when a better term is star system or planetary system.

Fig. 8.5 Admiral Adama (Edwards James Olmos) presides over the Combat Information Center aboard the battlestar *Galactica*. (Copyright © NBC Universal Television, David Eick Productions; Image credit: Photofest)

As mentioned above, in the reimagined *Battlestar Galactica* (Fig. 8.5), the ship is capable of making instantaneous faster-than-light jumps of up to 5 light years[19]. Over the course of the series, all of *Galactica*'s travels, along with her Rag Tag Fleet, are constrained within the small red circle in Fig. 8.4). The vast scale of the Universe, of one Galaxy (ours), shows why it makes little sense to talk about visitors "from another galaxy" or to say that Earth is "…far out in the Universe… in a galaxy much like our own."

8.3   A Star-Studded Gala(xy)

I don't know anything with certainty, but seeing the stars makes me dream.

Vincent van Gogh

Astronomers are like paparazzi. They take pictures of stars, always without their consent, and determine who is hot and who is not. Stars are classified their color or *spectral type*—which is a function of their temperature, and a re-

---

[19] This number was never stated explicitly in the series, but both writer Bradley Thompson and science advisor KRG arrived at this value through independent means.

flection of Wien's Law. Therefore, by observing the color of a star, astronomers instantly know its temperature, and a host of other parameters.

At the core of a star, the temperature reaches millions of Kelvins (that's also millions of degrees Celsius, since the 273 degree difference between the two temperature scales is negligible). The temperature is so high that atoms do not exist—collisions between atoms strip electrons from, or ionize, atoms at temperatures far lower. Rephrased, the matter in a stellar core is in the plasma state—a random assortment of electrons and atomic nuclei.

Atomic nuclei move so fast that their kinetic energy can overcome the electrostatic repulsion of their charges (recall that like charges repel, and all nuclei have positive charges). They slam together, sometimes stick, and create both new elements and enormous amounts of energy. The incredible pressure from the overlying mass of the star is what allows this process to continue, and is also why it is difficult to create fusion power on Earth—there is no way yet to confine a significant amount of plasma to allow it to undergo fusion in the same way as does a star. Through the process of nuclear fusion, our Sun converts 600 million tons of hydrogen to 596 million tons of helium every second—the remaining 4 million tons are converted into energy by $E = mc^2$. After a very long time (about a million years), that energy makes its way to the surface to be radiated as sunlight[20].

One tool astronomers use to characterize stars is the *Hertzsprung-Russell Diagram* (or simply the H-R diagram), seen in Fig. 8.6. In some sense you could look at the H-R diagram as Stefan's Law (hotter = brighter) plotted against Wien's Law (hotter = bluer): stars are positioned based upon both their temperature (or color) and total energy output. The horizontal axis represents surface temperature, which, in turn, dictates color. The vertical axis represents stellar energy output per second, and is a star's *luminosity*. The range of luminosities in Fig. 8.6 are based upon the luminosity of Sol; luminosities in the figure range from 10^{-5} (1/100,000th) to 10^6 (1,000,000) times that of Sol.

While the X-axis on most plots is displayed with the values increasing from left to right, on the H-R diagram the cooler red stars are to the right, and the hotter blue stars to the left. Different spectral classes range from class O for the hottest blue stars, to class M for the more common cool red stars. Astronomers have a fun mnemonic for remembering the progression (from hot to cool) O-B-A-F-G-K-M: "Oh Be A Fine Girl/Guy, Kiss Me." From cool to hot, it is, "Mickey Killed Goofy For A Body Organ."

---

[20] It is actually a fairly complex process, and the energy initially generated as gamma rays by the fusion process never actually reaches the surface of the star. It is converted into different forms in its path to the surface before it is radiated away as sunlight. Also, neutrinos created in fusion reaction zip through the star as if all that plasma wasn't there, which is why can be sure that fusion is still occurring today deep in the heart of our own Sun.

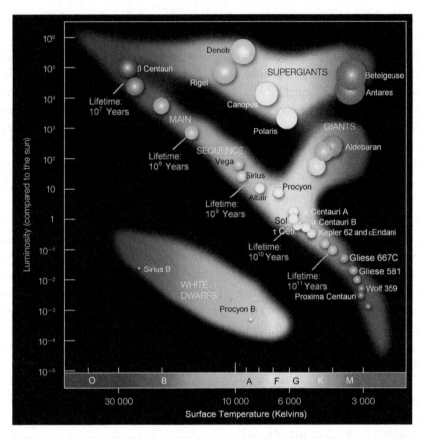

Fig. 8.6 The Hertzsprung-Russell Diagram with Main Sequence lifetimes. Sol will live another 5 billion years. (Figure credit: European Southern Observatory)

8.4 A Star is Born

Stars live the bulk of their lives on a sigmoidally-shaped region of the H-R diagram called the *Main Sequence* that runs from the upper left (hot, bright, and blue) to the lower right (cool, dim, and red). Stars begin their lives to the right of the M spectral class, and move onto the main sequence as they compress, heat up, and fusion begins. It turns out that the lone determinant of a star's life—its color, temperature, brightness, lifespan, and the method of its death[21]—is the amount of mass it had when it began its life as a star. Less massive stars are cooler, dimmer, live long lifetimes and their colors are towards

---

[21] Barring unusual circumstances, such as getting sucked into the supermassive black hole at the heart of most galaxies.

the red end of the visible spectrum. More massive stars are hotter, brighter, live short lifetimes, and tend towards the blue end of the spectrum.

For most stars on the main sequence, there is a linear correlation between their color and brightness. What that means is that if an astronomer measures accurately the color of a star, then the astronomer also knows the temperature and brightness of the star. Knowing how bright a star really is, and comparing that to how bright a star appears (increasing distance makes stars look dimmer) is one method astronomers determine distances to celestial objects, a technique called *main-sequence fitting*.

Sol is a G star. Within each spectral class there are 10 subdivisions, numbered 0–9, so to refine its classification further, our Sun is a G2 star. Whereas the Sun is often called a "yellow" star, the wavelength where it emits most of its radiation is actually in the yellowish-green portion of the spectrum. Science fiction (and a fair amount of popular science) has been efficient at propagating the myth Sol is "just an average star." It's true that depending upon how the X axis is presented (whether the spectral classes are equally-sized, or the temperature range is linear as in Fig. 8.6), Sol typically does fall close to the middle on the H-R diagram, which explains the tendency for people to say it is "average."

In the original *Cosmos*, even renowned planetary scientist Carl Sagan said, "There are some hundred billion galaxies, each with, on the average, a hundred billion stars, 10^{11} times 10^{11} equals 10^{22}, ten billion trillion. In the face of such overpowering numbers, what is the likelihood that only one ordinary star, the Sun, is accompanied by an inhabited planet? Why should we, tucked away in some forgotten corner of the cosmos, be so fortunate?"

There is a big difference however between an object whose classification falls at the midpoint between the end members of a spectrum, and an object that is *average*. That midpoint object would be average only if all classifications were equally likely. In fact, small M-type red dwarfs like Wolf 359 and Proxima Centauri are very numerous, comprising about 93 percent of known main-sequence stars, simply because they require less raw material (chiefly hydrogen) to form. There are a lot more orange K-class stars than there are Sol-type stars for the same reason. In fact, Sol is larger than over 95 percent of the main-sequence stars in our galaxy, with very large, and very hot, stars something of a rarity: large blue stars like β Centauri represent only one in 3,000,000 of the known stars.

Although it may seem counterintuitive, because they start out with so much more fuel "in the tank", so to speak, but larger stars actually live shorter lifetimes—in larger stars, a larger percentage of their bulk participates in the fusion process. Therefore, bigger stars expend their fuel much more rapidly.

The largest of stars live by the motto, "Live fast, die young, leave a good-lookin' black hole". Main-sequence stars in the upper left of the H-R diagram live, perhaps, a million years, while stars in the lower right may live trillions. The Universe is in fact not yet old enough for any M class red dwarf *ever* to have come to the end of its life on its own (though it is likely that some have met their demise by being too close to a companion star that exploded).

All stars, by definition, fuse hydrogen into helium, but bigger stars like Sol also fuse helium to make other elements like carbon, nitrogen, and oxygen. The process of creating new elements by fusing nuclei is called *nucleosynthesis*. The largest stars can synthesize rock-forming (hence terrestrial planet-forming), elements like silicon, aluminum, calcium, even nuclei as massive as iron, in their main-sequence lifetimes. Since iron is far from the most massive element known, from where did all the other elements, like copper, gold, silver, uranium, or lead arise? For that matter, how did these elements make their way out of stars to form planetary systems? For the answer, we examine how stars end their lives.

8.5 The Fault in Our Stars: Star Death

As stars age, they move off the main sequence. There are three other notable regions on the H-R diagram: one represents a stellar graveyard, the other two stellar hospices where stars go to live out the end of their days. Which region a star will migrate into, and how stars end their lives, depends entirely upon how much mass they had to begin with.

When the first red dwarfs finally die, they will, most likely, just simply stop fusing hydrogen and become a variant of what are known as brown dwarfs—objects that are intermediate between the largest planets and the smallest stars.

Stars nearer in mass to our own end their lives quite differently. Near the end of their lifetimes, as they can no longer fuse hydrogen as a primary power source, their outer layers balloon out. The stars become more luminous, or brighter, but are actually cooling down. They emit more light simply because of their enormous size, and they move off the main sequence. These stars are called *giants*, and are found up and to the right of the main sequence in the H-R diagram. Eventually these stars expel their outer shell, creating a cloud of gas called a *planetary nebula*. What is left behind is an extremely hot stellar core, called a white dwarf, below the main sequence in the H-R diagram. When our star enters its red giant phase, the outer layers will expand to beyond the orbit of Earth, frying the inner planets of the Solar System[22]. The resulting white dwarf will be about the same size as Earth, and will take trillions of years to cool down.

Approaching the end of their short lives, larger bluer stars swell to become *supergiants*, which can be found in the uppermost right region of the H-R diagram. While small stars can fuse hydrogen and helium into some of the lighter elements like carbon, nitrogen, and oxygen, larger stars can synthesize even heavier elements, all the way up to iron, which has a nucleus made of 26 protons and between 29 and 32 neutrons. Iron is in fact a star killer, a poison that eventually destroys the very thing that created it. The reason is because for elements less massive than iron, the fusion process generates energy. For elements more massive than iron, fusing nuclei consumes energy. So iron is effectively the dead-end of a star's fusion process. As iron builds up in the core of a star, the core becomes denser and denser as the amount of energy the star generates decreases. This renders the star unable to counter the crushing gravitational forces produced by its own tremendous mass. Eventually the gravitational overburden on the core becomes so great that electrons and protons are forced together to become neutrons in one huge catastrophic event. With no electrostatic repulsion to hold neutrons apart, they collapse to an incredibly dense state called neutronium[23]: a form of matter so dense that a teaspoon of the material would weigh several tons in Earth's gravity. A stellar core made entirely of neutrons is called a neutron star. If that core has enough mass, its gravity can be so strong that not even light can escape from it—then we refer to it as a black hole.

> The nitrogen in our DNA, the calcium in our teeth, the iron in our blood, the carbon in our apple pies, were made in the interiors of collapsing stars. We are made of starstuff.
>
> Carl Sagan, Planetary Scientist, *Cosmos*

This final collapse generates enormous amounts of energy in the form of gamma rays, plus a staggering amount of neutrinos, in one brief, intense, pulse. That pulse blows off the outer layers of the stars in a huge explosion known as a *supernova*. Supernova explosions release so much energy, that a single supernova can outshine the sum total of all the other stars in its galaxy for about a week[24]. The shockwave that propagates through the star to blow

---

[22] The second episode of the continuing tales of *Doctor Who*, "The End of the World," depicted this event.

[23] The term neutronium, sometimes neutrium, was popularized by science fiction, but was originally suggested in 1926 by German chemist Andreas von Antropoff for an element with "atomic number zero." In scientific literature it is, more commonly, called *neutron-degenerate matter*.

[24] A supernova whose light arrived at Earth in the year 1054 could be seen during the daytime. The remnant from that event can be seen with a good pair of binoculars if you look between the horns of Taurus, the bull. It is now called the Crab Nebula.

off its outer layers creates temperatures and pressures so high that, in this very brief event, every element heavier than iron is synthesized.

We see all of nature around us reflected in the life and death of stars. Plants take sunlight and basic nutrients from the ground, and build complex organic structures. Animals ingest plants adding further to the complexity of the web of life. In a similar fashion, the Big Bang created only hydrogen, helium, and a tiny bit of lithium—the raw material for stars—but in the early Universe, there were no planetary systems. The earliest stars processed hydrogen and helium into more massive elements. The expanding nebulae left in the aftermath of supernovae, were enriched in heavier elements, but prior to the first round of stellar deaths, planets, and things that could live on planets, were impossible. Throughout Nature, some things have to perish in order that others may live. We are star stuff, and it all started with a bang.

> There is in all things a pattern that is part of our universe. It has symmetry, elegance, and grace—those qualities you find always in that which the true artist captures.
>
> Frank Herbert *Dune*

Science Box: "Restarting" the Sun in *Sunshine*

> *Two last hopes are better than one.*
>
> Dr. Robert Capa, *Sunshine*

The 2007 UK film *Sunshine* is a gripping science fiction horror film, with the emphasis placed on the horror. In the film, our Sun is dying, and Earth is freezing. The spacecraft *Icarus* is dispatched with a stellar bomb—a huge device having the "mass of Manhattan Island"—in order to restart the Sun. All contact is lost with *Icarus*, and we join the drama seven years later with the crew of *Icarus* II en route to the Sun. Adhering faithfully to Chekhov's Gun[25], naturally, *Icarus* II encounters *Icarus*. Horror ensues. In the DVD commentary, director Danny Boyle says that in horror films, monsters traditionally attack from out of the darkness, and that his vision was to, instead, tell a story when a fiend attacks from out of the light.

---

[25] Chekhov's Gun is storytelling principle that says that every element in a narrative should be necessary and irreplaceable. Anton Chekhov, for whom the principle is named, said, "Remove everything that has no relevance to the story. If you say in the first chapter that there is a rifle hanging on the wall, in the second or third chapter it absolutely must go off. If it's not going to be fired, it shouldn't be hanging there." From Valentine T. Bill (1987) *Chekhov: The Silent Voice of Freedom*, Philosophical Library.

Fig. 8B.1 Mission scientist Robert Capa (Cillian Murphy) dons one of the heavily insulated and highly reflective space suits aboard *Icarus II* in *Sunshine* (2007). (Copyright © Fox Searchlight Pictures; Image credit: Photofest)

In reality, the typical lifetime of a star like Sol is 10 billion years, and our Sun is presently less than five billion years old. The Sun will not be dying any time soon. Say it was, though. Could a stellar bomb with the mass of Manhattan Island restart it?

The challenge comes from determining the definition of "the mass of Manhattan Island." Is that just the buildings? Does that include any of the ground or rock beneath? Since the film referred to a structure having the mass of Manhattan, we will make the assumption that it is just the buildings based upon the apparent size of the payload (Fig. 8B.1).

Manhattan has approximately 47,000 buildings[26] with an average height of 53 meters (about 17 stories)[27]. The survey that made the determination found significant variation in the building heights, so we will say that the average building is between 10 and 20 stories. The Empire State Building is 103 stories tall, and has a weight of 365,000 tons, which equates to a mass of 3.3×10^8 kg, or 3.2×10^6 kg/floor. So with 47,000 buildings, Manhattan has between 470,000 to 740,000 floors, for a total mass between 1.5 and 3.0 trillion kg.

Each second a typical G2 star like the Sun converts 600 millions tons of hydrogen into 596 million tons of helium—the remaining 4 million tons[28], or 3.9×10^9 kg—is converted to energy. The mass of the stellar trigger, then is between 385 and 770

---

[26] http://www.nytimes.com/2013/12/22/nyregion/calculating-the-number-of-windows-on-buildings-in-manhattan.html

[27] http://www.capitalnewyork.com/article/culture/2011/01/1189117/other-manhattan-project-thankless-quest-understand-new-yorks-atmosph?page = all

[28] Impressively, it is mentioned correctly in *Sunshine* that this much matter is converted to energy in the Sun every second.

times more massive than the amount of mass the Sun typically converts to energy every second, meaning that if it were entirely converted to energy by $E=mc^2$, then it would yield the same amount of energy that the Sun releases every 6 ½ to 13 minutes. That would be a pretty substantial kick of energy,

It is also an upper bound, and probably an overestimate by a factor of hundreds of thousands to millions. A lot of the interior of the device was shown to be empty space at the end of the film, so it clearly is not the case that the trigger is a storage device for antimatter, which would be the only way to get 100 % conversion of all the mass of the trigger to energy.

More importantly, Dr. Robert Capa (Cillian Murphy) says, "We've mined all Earth's fissile materials for this bomb, there's not going to be another payload. The one we carry is our last chance." The term "fissile material" clearly implies that the stellar trigger is just a really really large thermonuclear warhead. This patently says that the explosive yield will be orders of magnitude less than the full conversion of the mass of the device into energy. The audience also sees how the trigger is supposed to function, breaking into smaller units—with numerous small explosions rather than one large explosion (making it a multiple independently-targetable solar-entry vehicle, or a MISV, similar to a MIRV). Since every unit would require hardware to encase the fissile material, and to trigger its explosion, that means even less mass is dedicated to going BOOM.

Whittling away at the potential yield of this device, as described in *Sunshine*, clearly employing this stellar trigger with the "mass of Manhattan Island" to re-start a dying Sun is like using a firecracker to detonate a nuclear warhead. There is a lot to recommend *Sunshine*, and the film takes the viewer on a fun journey, but from a science perspective just suspend disbelief, and go along for the ride.

9
Shortcuts Through Time and Space

You might emerge somewhere else in space. Somewhen else in time.
Carl Sagan, Planetary Scientist

I believe our adventure through time has taken a most serious turn.
Ted Logan, *Bill and Ted's Excellent Adventure*

If time travel is possible, where are the tourists from the future?
Stephen Hawking, Theoretical Physicist, *A Brief History of Time*

It's one of the most exciting moments in any episode of *Star Trek: The Next Generation*: Captain Jean Luc Picard points forward, commands "Engage!"—and the stars stretch into blurred streaks as the *Enterprise* leaps to warp speed, its destination some unexplored star system with strange new worlds. Equally choice moments—the *Millenium Falcon* outrunning Imperial forces when Han Solo kicks in the hyperdrive, or the SG-1 team of explorers stepping through the shimmering horizon of a wormhole to find themselves walking on the surface of an alien world an instant later.

Each of these moments also represents one of the three most popular fictional techniques to travel at superluminal speeds, or faster-than-light (FTL)—namely warping space, ducking into a different "dimension," or connecting two distant regions of space so that traveling between them becomes manageable. In this chapter we'll be taking a close look at these methods on the basis of what we know about real physics—as well as the deep connection between FTL travel and another popular transportation technology, the time machine.

FTL travel is one of the oldest and most well-established tropes in both literary and screen science fiction. Its invention became a dramatic necessity owing to facts filtering out from the scientific community in the early twentieth century. One fact was that universe had a hard speed limit. One of the two postulates that Albert Einstein established in his 1905 Theory of Special Relativity[1] was that the speed of light was a universal constant, and no object

---

[1] Einstein, Albert (1905) On the Electrodynamics of Moving Bodies, *Annalen der Physik* 17: 891–921.

could travel faster than the speed of light through a vacuum[2], which zips along at just under 300,000 kilometers per second[3,4].

In addition to the invariance of the speed of light, the second postulate of Special Relativity says that there is no such thing as a *preferred reference frame*. What this means is that if an Earthly spacecraft is zooming past an alien planet, and there are observers on both, it is equally valid for the alien on the planet to believe it is at rest, and the spacecraft is moving past, or the human in the spaceship to consider the spacecraft to be at rest, and the planet zooming past. There is an old joke that says that a graduate student approaches Albert Einstein on a moving train and asks, "Excuse me, Professor Einstein, does New York stop at this train?" Importantly, this implies that there is no universal standard of being at rest, only being at rest relative to some other object.

Taken separately, these two parts of Special Relativity seem somewhat straightforward. Taken together, their implications are staggering. Say there are twins, call them Luke and Leia, moving towards one another, each in their own spacecraft. Leia fires a laser at Luke[5]. Both spacecraft will measure the speed of the laser to be c. No matter how fast the two are moving relative to one another, converging or departing, they will *always* measure the speed of light as c. What happens to make this true is that measurements that we tend to think of as invariant—distance and the rate at which time passes—will vary between observers to ensure that the speed of light is always c.

Why can't anything travel faster than light? What imposes the light speed limit is that, as an object moves faster and faster, it behaves as if its mass is also increasing. If we want to accelerate an object, say, a train carriage, we apply a force to it, pushing or pulling the carriage with an engine. The amount of acceleration produced is directly proportional to the force applied divided by the mass of the carriage, according to Newton's Second Law of Motion. This makes everyday sense: if we want something heavy to move at a given speed we need to exert more effort than to get something lightweight to move at that same speed.

In Newtonian physics, if it requires a certain force to accelerate, say, a train from 10 kilometers per hour to 60 kph, then applying the exact same amount of force to the train again will add another 50 kph to its speed, bringing it to 110 kph. The mass of the train is the same, the force is the same, so the acceleration is the same. In Einstein's Relativistic physics, applying the same force a second time will actually produce a train travelling a little *less* than 110 kph. This

---

[2] Light travels slower when passing through transparent materials, such as glass or water. This speed variation is actually the ultimate basis for how things like optical telescopes and microscopes work.

[3] Or over seven times around the equator of Earth each second.

[4] The explanation usually given for why the letter c is used to denote the speed of light is that c stands for "celeritas," the Latin word for "swiftness."

[5] Low power, of course, like a laser pointer.

is because in accelerating the train the first time, we have increased its energy, which means we have effectively increased its mass as per the famous $E = mc^2$; with more mass to push, we get less acceleration bang for our force buck.

For the kind of speeds a train can attain, this effect will never be noticeable, but then imagine a spaceship with a rocket engine constantly blazing away. As the speed increases, it becomes ever more difficult to maintain a constant acceleration, because the craft accelerates as if the mass is increasing as well. Particle physicists can measure this effect directly in accelerators like the Large Hadron Collider at CERN. The LHC is designed to speed protons up to a maximum of 99.9999991 % of light speed, at which point they behave as if their mass is increased by a factor of 7460. Even for subatomic particles, reaching exactly light speed would require infinite energy because, at that point, their mass acts as if it is infinite. This, naturally, makes accelerating *through* light speed flat-out impossible, and even accelerating objects—from a subatomic particle to a spaceship—to exactly the speed of light would require an infinite amount of energy. In the classic *Battlestar Galactica* episode "The Living Legend"[6] Commander Adama is in a position to use *Galactica* to support the Battlestar *Pegasus* in a major battle with the enemy Cylons. Adama assures her commander, Cain, "If you don't mind me using up half our fuel, I'll even see if I can't bring her [*Galactica*] up to light speed." It would take *all* of *Galactica*'s fuel to accelerate to even a large fraction of light speed, and reaching exactly the speed of light is almost certainly impossible. To go travel from point A to point B faster than light is going to require more than brute force.

In 1967, Columbia University physicist Gerald Feinberg published a paper on the possibility that there could exist particles whose minimum speed is that of light. He called these particles tachyons[7]. It did not take long for science fiction writers to latch onto the concept of tachyons as a way to enable FTL travel or communications, to the point where they have become a trope. They have been co-opted extensively in the *Star Trek* franchise, were mentioned in the film *K-Pax* (2001) in which Prot (Kevin Spacey) claims to be an alien from a binary star system in the constellation Lyra, whose journey to Earth was made possible by tachyons, and pop up in the technobabble of numerous shows and movies.

To date there is no evidence that tachyons exist—primarily because, traveling faster than light, they would travel backwards in time. In September 2011, CERN incorrectly reported that a subatomic particle, a tau neutrino, travelled faster than the speed of light. The researchers later discovered that a hardware fault had led to this erroneous conclusion. The laws of relativity have not been broken, that we know of, yet.

---

[6] Voted the best episode of the series in almost every poll.
[7] Feinberg G (1967) Possibility of Faster-Than-Light Particles. *Physical Review* 159(5): 1089–1105.

Why are we so desperate to beat light speed? Because another fact that throws a wet blanket on the promise of interstellar travel is that the universe is *really* big. An excellent example illustrating the immensity of interplanetary and interstellar distances is on the National Mall in Washington, D.C. Starting in front of the Smithsonian Air and Space Museum, running along Jefferson Drive, the Voyage Model of the Solar System is a 1:10,000,000,000 (one to ten billion) scale model of the Sun, planets, and Pluto. At that scale, Sol is about the size of a small cantaloupe. Earth is about 50 feet away, and is smaller than the size of a pinhead. At 255 feet from Sol, Jupiter is marble-sized. At the far end of the Air and Space Museum, Uranus is just short of 950 feet away. Across 7th Street, at each end of the Arts and Industries Building, are Neptune and Pluto. Pluto is a little over 1910 feet from the Sun (Fig. 9.1).

> Space is big. Really big. You just won't believe how vastly hugely mind-bogglingly big it is. I mean, you may think it's a long way down the road to the chemist, but that's just peanuts to space.
>
> Douglas Adams, *The Hitch-Hiker's Guide to the Galaxy*

Using the same scale let's imagine adding a stellar neighbor. The nearest star to Sol is the trinary star system Alpha Centauri, which might be represented by a slightly larger cantaloupe than that representing Sol (Alpha Centauri A) in mutual orbit with a large grapefruit (Alpha Centauri B), with a cherry (Proxima Centauri) in orbit around them both. All three stars would be located in San Francisco's Golden Gate Park, over 2500 miles away! At this scale, light would cross the distance at about one inch per second, or the rate an ant walks, taking about 4.3 years to traverse the country.

Astronomers had measured the distance to some of the stars in the Milky Way as far back as the nineteenth century, but in 1924 Edwin Hubble blew the lid off by announcing that not only did other galaxies exist outside our own, but that the closest large Galaxy—M31 in the constellation Andromeda, known also as the Andromeda Galaxy—was a staggering 2.5 million light years away. (Recall that one light year is the distance light travels in one year: about 9 ½ trillion kilometers or 6 trillion miles in real space.)

Cranking up a spaceship's velocity for a lightspeed journey to Andromeda would still mean a 5-million-year round trip. Even for those of us that enjoyed the carefully measured pacing of 1969's *2001: A Space Odyssey*[8], a screenplay based upon that journey would make for a very boring film. Five million years is a long time to keep a spaceship in good running order. Five million

---

[8] SAC enjoys the pacing of *2001*; KRG is a heretic.

Fig. 9.1   The Sun for the Voyage model of the Solar System in Washington, D.C. Pluto is about 1900 feet to your left. (Photo by Michelle Thaller)

years ago was roughly the time our ancestors went their evolutionarily separate way from today's chimpanzees, so you have to wonder what species would be around to greet the returning expedition. So, at first glance, Special Relativity appears to be a total downer when it comes to cool interstellar adventures, and it completely obviates intergalactic drama. Thanks for nothing, Einstein!

In actuality, Uncle Albert left us with an interesting loophole: time dilation. Time dilation means that as our twins Luke and Leia travel ever faster relative to one another, each one would see the clock aboard the other's spacecraft moving ever slower. It is important to note that when each twin looks at the clock aboard their own spaceship, each will still perceive one second to occupy the same amount of time as it always has. (It is also important to note that while, technically, time dilation happens for *all* objects moving relative to one another, it's only readily noticeable at speeds that are a significant fraction, above 10 % or so, of *c*.)

If an astronaut could travel at very close to light speed, then she could blast off from Earth towards a star, say, a few hundred light years away. From Earth, we would see her spaceship take centuries to reach the star. Depending upon how fast the spacecraft traveled, the journey might only take a few decades for our astronaut, maybe even just a few years. Then she could turn around, return home, and, once again, compress the apparent travel time. She would find that, indeed, centuries had passed in Earth since she left, and every one she once knew was long dead.

The existence of relativistic time dilation is not theoretical. It was first confirmed experimentally by physicists Rossi and Hall back in 1941. When cosmic radiation interacts with the rarefied atoms of Earth's upper atmosphere, they create a shower of subatomic particles, some of which are particles known as *muons*. On average, muons decay into less massive subatomic particles in slightly over 2 millionths of a second[9]. Muons travel at very high speeds, but even photons, traveling at the speed of light, would travel only 2000 feet (600 meters) in 2 microseconds. Yet numerous muons reach Earth's surface. This is due to relativity: because muons have mass and travel at nearly the speed of light, their lifetimes are extended due to time dilation, and a significant fraction of them live long enough to impinge upon Earth. Subsequent experiments using atomic clocks have repeatedly verified the effect.

This consequence of special relativity underlies the premise of a number of science fiction movies, perhaps most famously 1968's *Planet of the Apes* (spoiler alert!): after a long near-lightspeed journey, astronauts crash land on what they believe is an alien planet—one that is actually Earth of the distant future. Time dilation is also the reason offered why the alien abductees released at the end of *Close Encounters of the Third Kind* (1977) appear to have aged little, despite the passage of decades on Earth in their absence.

Still, even a relativistic travel trip time of just a few years poses challenges. Many movies that try to convey the notion of long-duration space missions but don't want the bother of worrying about crew mental stability or depletion of resources, avail themselves of another now-classic science fiction trope: suspended animation, where crewmembers are placed into pods, their bodies put into some form of stasis or hibernation that prevents or slows aging, and who are then awakened upon arrival at their destination[10]. *Alien* (1979), *Aliens* (1986), and *Prometheus* (2012) leveraged this idea particularly well in their narratives, using it deftly to establish just how isolated and far from rescue the characters were. Dozens of other examples abound, including *Lost in Space* (1998), *Planet of the Apes, Outland* (1981), *2010: The Year We Make Contact* (1984), *Avatar* (2009), *Interstellar* (2014), and as a key piece of the backstory of *Defiance* (2013-) in which the action takes place on Earth after aliens onboard an armada of spacecraft wake up after a long voyage across the Galaxy. Even the Golgafrincham ark from *The Hitch-hiker's Guide to the Galaxy* employed a form of suspended animation.

---

[9] That's 2 microseconds, or 2 μs.

[10] Suspended animation is also a long way off, but like FTL travel, has had some promising recent developments.

If suspended animation or relativistic travel isn't an option, then a few possibilities remain. One approach left for crossing interstellar distances is to construct so-called generation ships, in which the original spacecraft crew live their lives, have children, and die. To avoid inbreeding, these ships must be big enough to accommodate hundreds of passengers, plus the farms, factories and recycling facilities required to keep everyone alive, and enough resources to allow the process to keep recurring, while people living their lives in the normal fashion (or as normal as possible), until reaching their destination. With any hope, the individuals who complete the journey will be of the same species as the original crew. This theme has been explored many times in screen science fiction, in *Star Trek* (TOS and *Voyager*), *Space: 1999*, *Firefly* (2002), and in great detail in *Ascension* (2014–).

There are a number of ways to extract drama from a premise in which a crew expects little to happen for generations, but a typical approach is to have the passengers forget that they are actually on spaceship, as in the ill-fated TV series *The Starlost* (1973–1974). While capable of generating interesting stories, the approaches above do place significant limits on the kind of storytelling that can be done with interstellar travel. It makes it very difficult to have the rambunctious high-species-count, high-planet-count, high-spaceship-count settings of the *Star Trek*, *Babylon 5*, *Stargate*, or *Doctor Who* franchises, for example.

So out of sheer narrative necessity, early science fiction writers invented FTL travel. No need to spend decades or centuries getting from A to B: just turn on the FTL machine onboard your characters' spaceship and arrive either instantaneously or at least after a journey taking no more time than that of a nineteenth century sea voyage. This narrative convenience is likely why the 2013 movie adaptation of the 1985 novel *Ender's Game* incorporated FTL travel, despite the fact that the implications of slower-than-light relativistic travel underlies several key plot elements in the novel.

For many productions, then and now, little attention gets paid on screen to the actual mechanism used to bypass Einstein's universal speed limit—a "hyperdrive" or "FTL drive" might be mentioned in passing, but little more. Because superluminal travel is such an established trope, audiences are rarely bothered by this lack of specificity, much as they usually never question how the artificial gravity is implemented or how the phenomenal energy output of a blaster is produced by something small enough to grip in one hand.

Without some practical limits though, FTL travel could rapidly siphon off dramatic tension—if the characters can escape to another star system every time they are in danger, how can they ever be in jeopardy? For some shows, the limits don't principally arise from the technology itself, but are externally imposed. For example, in both the old and new *Battlestar Galactica*, the two biggest limits on travel are from coordinating the movements of an entire fleet, and the fact that most of the time the fleet is in completely unknown territy without a clear destination, and so must scout its way forward to locate resources, navigate hazards, or reconnoiter for hostile enemy craft.

Other movies and series choose to spend a little more time establishing a few of the broad principles of their transportation technology, thereby establishing operational demands and limits that can be mined for dramatic purposes. These include things such as the rule in the *Stargate* franchise that a wormhole generated by one of the gates can only be kept open for 38 minutes: without this rule, expeditionary teams could be in constant contact with Earth, summoning backup whenever needed, while also preventing incoming hostile forces from accessing the local stargate.

The start of this chapter highlighted three different FTL technologies, each of which represents one of the three broad categories into which fiction FTL travel falls: warping space, passage through an alternate realm (ala "hyperspace"), or traversing a wormhole. Let's explore each one in turn.

Science Box: Putting Einstein in the Pilot's Seat

In the *Back to the Future* movies, the DeLorean time machine works only if it has just over a gigawatt[11] of energy on hand *and* is travelling at 88 miles per hour. In real life, *any* relative motion between two observers will mean that each perceives passage through time differently; the faster you go, the slower time will run for you relative to an outside observer at rest: this is the time dilation effect, and in theory it could allow you to make an interstellar journey that would take centuries from the point of view of people of Earth, but for a crew might require a scant few years.

> *If you're gonna build a time machine into a car, why not do it with some style?*
>
> Doc Brown, *Back to the Future*

However, to get any sort of appreciable dilation, let alone appreciable, you must be travelling way faster than 88 mph. To calculate just how much, we're going to pull an important equation out of Einstein's Theory of Special Relativity. Special Relativity has two parts, the first being that the speed of light in a vacuum is an invariant. No matter how fast two objects move relative to one another, they will always measure the same speed for light. What this means is that metrics that we tend to think of as constant, like distance and the rate of passage of time, will actually change between moving observers such that they will both measure the same value for the speed of light. We can calculate these changes using something called the *Lorentz factor*[12], γ:

$$\gamma = \frac{1}{\sqrt{1-\frac{v^2}{c^2}}} = \frac{1}{\sqrt{1-\left(\frac{v}{c}\right)^2}}$$

[11] Doc Brown pronounces it "jigga watt" when the pronunciation is "giga watt" with a hard "g".
[12] Eagle-eyed watchers of (2006–2012) may have noticed that the Lorentz factor equation is one of the equations Sheriff Carter is forced to memorize in the time-loop episode "I Do Over"—in his role as consultant to the show, KRG supplied the math.

Here, v stands for the relative velocity between, say, a speeding spaceship and Earth, while c is the speed of light—300,000 kilometers per second. To calculate how much slower time passes aboard the spaceship, we multiply the elapsed time on Earth by the Lorentz factor. If we were to start two clocks, one aboard the spacecraft, and one on Earth, the relative passage of time would be

$$\Delta t_{SC} = \gamma \Delta t_{Earth},$$

where Δt_{SC} is the elapsed time on the spacecraft, and $\gamma \Delta t_{Earth}$ the elapsed time on Earth. Let's try a few test values for the spaceship velocity to explore how time dilation changes as we near the speed of light.

Start with the top speed humans have travelled in space to date, 11 km/s (reached during the return to Earth by the Apollo 10 mission), gives us:

or,

$$\gamma = \frac{1}{\sqrt{1 - \frac{\left(11^{km}/_s\right)^2}{\left(3 \times 10^{5km}/_s\right)^2}}},$$

finally,

$$\gamma = \frac{1}{\sqrt{1 - 0.00000000134}},$$

$$\gamma = 1.00000000067.$$

This means that for every 1.00000000067 seconds that pass on Earth, one second passes aboard the spaceship. So if one year—31,557,600 seconds—passes on Earth, then on the spaceship the year will be about 200 hundredths of a second shorter. That difference may be of academic concern, but not a practical one.

Ramp the spacecraft speed up to 30 percent of lightspeed, or 0.3 c. Using fractional values of the speed of light is convenient because we no longer have to convert to km/s and back as v/c is simply 0.3, meaning:

which becomes

$$\gamma = \frac{1}{\sqrt{1 - \left(\frac{v}{c}\right)^2}},$$

and,

$$\gamma = \frac{1}{\sqrt{1 - (0.3)^2}},$$

$$\gamma = 1.048.$$

Now, a year of Earth time is equal to 365/1.048 days, or 348.4 days, aboard the spaceship. That's certainly a noticeable difference, but not one that's going allow a crew to travel to a star system 100 light years from Earth.

Crank up the speed again to 90 percent of lightspeed—now $\gamma = 2.294$, making 365 Earth days equal to 159 spaceship days. Still not enough, so let's go to 99 percent lightspeed ($\gamma = 7.09$): now one Earth year ticks past in just over 51 days aboard the spaceship.

We're getting there, but still not enough, so lets really put the pedal down and go to 99.9999 % lightspeed, where $\gamma = 707.1$, meaning every 12.38 hours on board the spacecraft, a year passes on Earth. A century on Earth will pass in 3.4 years for those aboard the spacecraft. Ignoring any time required to accelerate and decelerate, a crew could go, have a little over three years to explore the star system 100 light years away, return, and while 10 years will have passed for the crew, 203.2 years (200 spent at near light speed, plus the time spent investigating the star system) years will have passed on Earth.

There's a "gotcha," though[13]. Recall that there are two elements to Einstein's Special Relativity. As the spacecraft speeds away from Earth, it is equally valid for the spacecraft crew to view the situation as Earth speeding away from them. In that situation, whose clock is running slower, the one aboard the spacecraft, or the one on Earth? Depending upon the location of the observer, it could be either, but Relativity also provides us a way to solve this ambiguity.

In addition to the rate of the passage of time, another metric that changes based upon relative motion is the concept of length. If one could measure a meter stick as it passed by, moving at a large fraction of the speed of light, it would appear shorter than it would at rest. This phenomenon is called Lorentz contraction, and an object's length as a function of its velocity is:

$$L_{\text{moving}} = \frac{L_{\text{rest}}}{\gamma}$$

Where L_{moving} is the length of the object travelling at a velocity v, and L_{rest} is its rest length. Not only does this work for the length of physical objects, it works for distances between stars. So

$$L_{\text{SC}} = \frac{L_{\text{Earth}}}{\gamma},$$

where L_{SC} is the travel distance measured in the moving spacecraft, and L_{Earth} the distance to the target star measured before the spacecraft began moving. If the distance to our target star is, again, 100 light years, and $\gamma = 707.1$ for a relative speed of 99.9999 c, then the distance between the start and finish of the journey becomes

$$L_{\text{SC}} = \frac{100 \text{ LY}}{707.1} = 0.14 \text{ LY},$$

which, again, is just over 51 days. In this instance, it doesn't matter if the spacecraft or the target star is considered to be the rest frame, the distance is Lorentz contracted by their relative motion and, still, over 200 years pass back on Earth during the journey.

Two centuries may seem like an impractically long time for a future Earth civilization to wait for the fruits of stellar exploration, but consider that it took European powers 300 years to finally complete an overland expedition from the East to the West coasts of North America.

---

[13] Isn't there always?

9.1 Let's Do the Space Warp Again

Now we have a sense of why the light speed limit exists, let us examine ideas for getting beyond it, starting with warp drives. Warp drives were something of a happy accident for the *Star Trek* franchise, in that the details were never clearly established in the original series, but, in 1994, the name inspired physicist Miguel Alcubierre to search for an actual scheme that permits an FTL system based upon warping space.

The basic principle of Alcubierre's warp drive is that although it is forbidden for an object to travel faster than the speed of light, all bets are off when it comes to the fabric of spacetime. That can seem like a confusing statement, but Einstein's General Relativity tells us that space has a fabric that can stretch and bend and warp, and it can do so at faster than the speed of light while, at the same time, not violating the laws or Relativity—which we explored previously using the analogy of a rope and two ants.

Just as the Universe was able to expand faster than the speed of light after the first tiny fraction of a second after the Big Bang, Alcubierre's warp drive similarly exploits the malleability of spacetime. In principle, it creates a bubble of distorted space around a spaceship. Space in the direction of travel of the spaceship is contracted, while space behind the spaceship is expanded. The spaceship moves forward through the higher "density" region of space, while the ship never exceeds the light speed limit *within* the bubble. This enables FTL travel while violating no rules.

Unfortunately nobody is rushing to build a working prototype because there are serious practical problems. From 1996 to 2002 NASA ran the Breakthrough Propulsion Physics Program at the Lewis Research Center[14] near Cleveland. The goal of the program was not to develop a working propulsion system, but rather to search for breakthroughs in physics that would circumvent known technical barriers to space drives and FTL travel. Alcubierre's work was one of the avenues this program explored.

In the case of the Alcubierre drive, one problem is that generating the warp bubble requires a field of negative energy surrounding the space craft, and there is no way at present to produce such a field. The Alcubierre drive also requires a *lot* of energy, far more than even the antimatter core on board *Enterprise* (any *Enterprise*) can produce. Theoreticians have been working to solve these problems, and, for example, have managed to reduce the original energy requirements from the ludicrously absurd to the insanely impractical, but there is still a very long way to go[15].

---

[14] Some former members of the program now run the private Tau Zero Foundation, which has similar aims.
[15] NASA used to host a nice web site called "Warp Drive, When?" that was a companion to the Breakthrough Propulsion Physics program. There is still a great annotated bibliography of the relevant physics of FTL travel here: http://www.nasa.gov/centers/glenn/technology/warp/bibliog.html.

> (Regarding the warp drive part of the set) I'm working on that.
>
> Stephen Hawking, while filming *Star Trek: The Next Generation*

A related concept can be seen in the various adaptations of *Dune* (the 1984 film and the 2000 miniseries). In these, Guild navigators on vast Heighliner spacecraft fold space ahead of them so that *there* becomes *here* and 20 kilometers of spaceship simply disappears from one star system and reappears in another. Because folding space is a tremendously complex operation and computers have been banned in the wake of a particularly messy robot uprising, navigators must ingest the spice melange in order to give them the prescience required to sense the correct way to fold space to ensure the safest route. This makes melange critical to interstellar trade, and since it can be harvested on only one planet in the galaxy, spice is the primary economic driver for the conflict in the novel *Dune*, as well as many of its sequels[16].

9.2 Jump To Another Realm

Another way in which FTL travel is frequently depicted is for a craft to enter an alternate realm, often known as hyperspace. This term is used very loosely in science fiction, but for our purposes "hyperspace" implies a different plane of existence—a different universe, for example—somehow coexisting with our own. This plane of existence maps to our universe in such a way that if a spaceship shifts into hyperspace, travels some distance, and shifts back, it will appear in a vastly different place in our universe (This use of the term "hyperspace" was coined by John W. Campbell in a 1931 issue of *Amazing Stories*.).

If physical laws in hyperspace were the same as in our universe, it would not permit FTL travel, as the same speed of light restrictions would apply (although it might be useful for other things, like escaping from a locked room, if the room's walls existed in one universe but not the other. The characters on *Fringe* (2008–2013) use a similar dodge on occasion, shifting between this universe and a parallel one to avoid various obstacles.) Instead, physical laws in hyperspace are usually specified as being different in key respects, although not so different that they impede keeping a spaceship in working order. One approach is to depict hyperspace as a place where the speed of light is the *lowest* possible speed an object can travel, similar to the rules governing tachyons.

---

[16] This is also the FTL method most similar to that depicted in the reimagined *Battlestar Galactica*, and as described in the film *Event Horizon* (1997).

Fig. 9.2 In the *Babylon 5* universe, larger vessels could enter hyperspace by open-ing a jump point. Smaller vessels required the assistance of an external jump gate. (Copyright © Warner Brothers Television. Image credit: Photofest)

After entering hyperspace, a spacecraft can zip along at some multiple of light speed and then reenter our universe far from the starting point.

Another approach is to have short distances in hyperspace correspond to large distances in our universe, so that, say, 1 meter in hyperspace equals 100 kilometers in normal space. So travelling at 0.01 percent of light speed in hyperspace—which translates to a very doable 30 kilometers per second, less than twice the top speed of the *Voyager 1* probe—would be the equivalent of travelling at 10 times the speed of light in our universe, meaning that a space-craft could cross the 4.4 light years between Earth and the Alpha Centauri star system in less than six months[17].

This different distance idea was the tack taken by the TV show *Babylon 5* (1993–1998), where hyperspace is depicted a formless void where it's easy to get lost without navigational assistance from some form of beacon. A lot of energy is required to shift in and out of hyperspace, so most smaller space-craft must rely on a network of jump gates to open windows between normal space and hyperspace for them. In the *B5* universe, larger craft (i.e., your ba-sic aircraft-carriers-in-space) can open their own hyperspace windows called jump points (Fig. 9.2).

In *Star Wars: A New Hope*, Han Solo brags that his ship, the *Millennium Falcon* is so fast that it is, "…the ship that made the Kessel run in less than

---

[17] *Minecraft* players will recognize the similar mechanic that is employed for mapping travel in The Nether to the normal world.

12 parsecs." This statement can be interpreted in two different ways, and has been a subject of ongoing debate (in addition to who shot first) among *Star Wars* fans for decades[18]. A parsec is a unit of distance (3.26 light years), but, in context, Solo implies that it is a measure of time. This could mean that, like "light year" is often confused as a measure of time, Solo (or the screenwriter), was similarly confused. It could mean that his statement was B.S.: note that Obi-Wan Kenobi rolls his eyes at the boast.

This, alternately, could imply that the drive of the *Millennium Falcon* uses a marriage of technologies. Viewers learn early on that the *Falcon* travels through hyperspace to achieve FTL. Does the ship additionally *fold* hyperspace? Are there different levels of hyperspace? In that universe is the machismo of one's FTL drive measured by which one can bring distant points together the closest? Most importantly, the distance of one parsec is based upon the size of Earth's orbit around Sol. Why is a smuggler "in a Galaxy far, far away" even alluding to Earth in the first place?

There is some reason to believe that parallel universes, alternate realms, exist. For now, though, we have no reason to believe that the physical laws in other universes are different than in our own. Nobody is saying that could not be the case, either. For the time being, hyperspace is an alternate realm within science fiction alone.

9.3 Wormholes

The third big idea for FTL travel is to traverse wormholes. The term wormhole was first coined by physicist John Wheeler in 1957 to describe shortcuts through spacetime that result from mathematical solutions to Einstein's equations of General Relativity. Wormholes share similarities with warp drives, in that they rely upon the fact that in General Relativity space can be distorted in odd ways. In the case of a wormhole, distortions take a static point-to-point form rather than the moving bubble of the Alcubierre drive. Although many such solutions have been discovered over the years, it took until 1988 for theoretical physicists to find equations for wormholes that were consistent with General Relativity and which could be both stable enough and large enough for a spacecraft to traverse.

Typically, the way to imagine a wormhole is to start by ignoring one of the three dimensions of space and representing space as a flat sheet. Normally to

---

[18] This is the subject of a retcon attempt in a novel by A.C. Crispin. "Retcon" stands for retroactive continuity, a way of correcting a mistake *ex post facto*. In the novel the Kessel run was dangerous, but the *Falcon* was so fast that it could skirt a field of black holes that other ships could not to shorten the distance. As retcons go, we've seen worse.

get from one point to another means following a path along the surface of the sheet. A wormhole is like a tube rising up out of the sheet and connecting two distant points, so that someone travelling between the two points has a choice of paths to follow, one along the sheet and another along the tube.

We've all seen this kind of depiction, usually with a nice arrangement of grid lines, but it's important to realize that it can lead your intuition astray in an important way. In the flat sheet version, the path along the wormhole is always *longer* than the direct route along the sheet. Critically, in a "real" wormhole (*real* is in quotes because, after all, we haven't built one yet) the length of the wormhole can be arbitrarily small. Unfortunately, there's simply no way to visualize this with a sheet without bending the sheet over through the third dimension, so that the distance along the sheet is longer than the "direct" distance between the points, which is a bit of cheat: it's just one of those four-dimensional things.

It's also worth pointing out that the 2-D/3-D way of illustrating wormholes has often led to the depiction of the open end, or *throat*, of a wormhole as either a whirlpool, or a flat, circular, portal, as in the *Stargate* franchise. In fact— as both carefully explained by one of the characters and beautifully shown on screen in *Interstellar*—the throat of a wormhole is actually a sphere[19].

> I find that the more you stick to facts, and the more you have to stay clean, the better the story is. Because when you're making stuff up, and you can cut those corners, it ends up getting cheesy.
>
> Lynda Obst, Producer, *Interstellar* and *Contact*

So one end of a wormhole could be at Earth, while the other end could be in the Alpha Centauri system, all the while keeping the length of the wormhole at a few centimeters. Stepping through the wormhole could be done at gentle stroll, so there is no violation of the light speed limit within the wormhole, yet in a single step you have crossed 4.4 light years.

Like the Alcubierre drive, constructing a real wormhole requires conjuring up negative energy, but instead of a moving bubble, we just need to keep a fixed region of space filled with negative energy. There is actually a known method to do this on a small scale via the so-called *Casimir Effect*, which is one of the weirder consequences of quantum mechanics. In a nutshell, quantum mechanics tells us that there's no such thing as completely empty space

---

[19] *Interstellar*'s spectacular depiction has probably done a lot to kill off the misconception of wormhole throats as flat circles, but it wasn't the first to do so: *Primeval* (2007-2011) used space-time "anomalies" to move in time, depicting them as glowing mostly-spherical intrusions.

if you look closely enough and quickly enough: instead there's an endless sea of *virtual particles* winking in and out of existence with a distribution of energies. Essentially, each virtual particle borrows a bit of energy from the vacuum from which it is created, exists for a moment, and then pays back the loan to the vacuum when it vanishes. Only *on average* does space look empty and devoid of energy.

Place two conducting plates close to one another, and they will exclude some of the virtual particles. This creates a region of space with fewer particles, thus less energy, than the surrounding space—if you had a universal energy meter that read zero in normal space, it would read negative between the two plates, and, in theory this effect could be used to make and stabilize a wormhole. (The fact that the hardest part of making a wormhole is preventing it from collapsing the moment a single particle enters is reflected in the premise of *Star Trek: Deep Space Nine* (1993–1999), where the rarity of finding a stable wormhole between two distant sectors of the Milky Way is responsible for an otherwise backwater star system's high strategic value to several major powers.)

Before you start investing in shiny metal plates, you should know that the Casimir effect is really only appreciable if the plates are a very small distance apart, so using the effect to build a wormhole big enough for a person to walk through is an unsolved challenge.

As well as the aforementioned *Deep Space Nine*—where, interestingly, a narrative choice was made to put the wormhole essentially off-limits for much of the show's run— wormholes feature frequently in the screen science fiction of recent decades, appearing in roles major and minor in shows such as *Sliders* (1995–2000), *Farscape* (1999–2003), and *Fringe*, and, most recently, in *Interstellar*.

The franchise that is far and away most identified with wormholes is the sprawling *Stargate* franchise—*Stargate* (1994); *Stargate SG-1* (1997–2007); *Stargate Infinity* (2002–2003); *Stargate Atlantis* (2004–2009); and *Stargate Universe* (2009–2011). The show deserves a lot of credit for slowly building up a plausible body of "wormhole physics" to go with their wormholes while trying as much as possible to maintain internal consistency. Of course, some things were fudged *vis a vis* actual theoretical physics, but even then not *too* badly.

One example of such a *Stargate* fudge is that, in real theoretical models of wormholes, you can't just pick another point in the universe and construct a wormhole to that point on demand. Instead, you have to construct both ends of the wormhole close by in space and then move one mouth of the wormhole at sub-light speeds to the desired destination. Only once the wormhole is in place can you then can step between worlds. If the wormhole completely collapses at any time then the link is permanently severed and a new wormhole must be constructed and towed out (Fig. 9.3).

Fig. 9.3 What awaits on the other side of the gate? Colonel Jack O'Neill (Richard Dean Anderson) and Jaffa Teal'c (Christopher Judge) are at the ready as they step onto a planet in *Stargate SG-1*. (Copyright © MGM Worldwide Television Productions. Image credit: Photofest.)

Of course this means having one pre-existing wormhole for every star system you might want to visit. This could rapidly become unwieldy, both from the production and narrative points of view, so the creators of the Stargate television series struck a compromise: the heroes would use just a single wormhole device, but one that can only connect to other stargates within a pre-existing network that had been built in the distant past by a mysterious alien race.

Stargate explored other interesting aspects of wormhole physics, most interestingly their temporal effects. In one *SG-1* episode ("A Matter of Time") the team connects to a stargate on a world orbiting a new black hole, which means one mouth of the wormhole is in a high gravitational field. This is a problem for them because in Einstein's Theory of General Relativity, gravity produces time dilation, just like travelling through space at high speed. More gravity means a slower clock. Because the wormhole links two places in space, the time dilation effects of the black hole's gravity begins to manifest themselves on the Earth side of the wormhole, with *SG-1*'s headquarters rapidly falling more and more out of sync with the rest of Earth as the team tries to disengage the wormhole.

> The characters in the pilot we wrote *do* time travel—they travel one second into the future every second, the way we all do.
>
> Zack Stentz, Writer/Producer

Fig. 9.4 Explorers Brand (Anne Hathaway) and Doyle (Wes Bentley) visit a water-covered planet orbiting a black hole in *Interstellar* (2014). The crew must deal with an environment exhibiting two features of existence deep within a powerful gravity well: immense tidal forces and time dilation. (Copyright © Legendary Pictures, Lynda Obst Productions, Paramount Pictures, Syncopy. Image credit: MovieStillsSB.com.)

Time dilation due to the high gravity surrounding a black hole was also a key plot point in *Interstellar*, where an already lengthy space mission gets even longer (from the point of view of those left behind) when members of the crew visit a planet orbiting close to a black hole (Fig. 9.4). This gravitation dilation effect is real, and has been measured directly using atomic clocks flying aboard passenger airlines and spacecraft[20]. Because these craft are farther away from Earth when they are in the sky than on the ground, they are in a slightly weaker gravitational field. Consequently they experience less of a time dilation effect, so time runs a little slower on the ground than in the air. This means that every time you take a flight you arrive at your destination a few nanoseconds older than you would have been if you'd stayed on the surface.

In several *Stargate SG-1* episodes, wormholes are also used to travel through time as well as space, which brings us to a device that rivals the FTL drive for popularity in screen science fiction: the time machine. This happens because space is really properly viewed a four-dimensional continuum, in which the three dimensions of space are inextricably bound with the fourth dimension of time, called spacetime. (It's not for nothing that the creators of *Doctor Who* in 1963 put the eponymous hero in a time machine that doubled as a spaceship and called it the Tardis for *Time And Relative Dimension In Space*.) Every object that moves through space also moves through time. Critically, as

---

[20] Every time you get a GPS reading, there has been corrections to account for (1) the spacecraft velocity, and (2) the difference in gravitation between you and the satellites in space. This is true GPS, not the faux GPS often used by modern smartphones.

we've seen with time dilation, *how* an object moves through space affects how it moves through time as well.

The time dilation effect we discussed earlier can be thought of as a one-way time machine, allowing us to travel into the far future as in *Planet of the Apes*, but what about "true" time travel—the ability to travel into the past? It turns out that that there are theoretical paths through space which *do* lead back through time. These paths are called *closed time-like curves* or *CTCs*. When CTCs were first discovered lurking within the equations of Relativity theory, they were dismissed as a mathematical curiosity, since it was originally believed that the entire Universe would have to be rotating for them to have any meaning in physical reality.

Attitudes to CTCs changed in 1974 when a physicist named Frank Tipler knocked out a scientific paper catchily entitled "Rotating Cylinders and the Possibility of Global Causality Violation." In it he proposed that the gravitational field produced by a massive spinning cylinder would warp space so severely that a spacecraft travelling around the cylinder could follow a CTC.

Tipler's cylinder's had several problems, not least that its mathematics was built around the basis of an infinitely long cylinder. It was Stephen Hawking who really killed off the cylinder by finding a fatal technical flaw in the theory. None the less, by that time dreaming up various theoretical time machines consistent with the known laws of physics had become something of a parlor game for physicists.

The most well-known of these putative time machines was developed by Kip Thorne[21], Mike Morris, and Ulvi Yurtsever in a paper published in 1988. The first step in their time machine assembly process is to build a short wormhole. Initially the two mouths of the wormhole start off close together, but then you grab one mouth, accelerate it to near light speed, fly out into space some distance, and then bring the mouth back to the starting point. Thanks to time dilation, the mouth that is accelerated has experienced the passage of less time than the mouth that stayed put. Stepping from the "older" mouth to the "younger" mouth would allow you move backward through time. As an alternative to a light speed journey, dipping one end of the wormhole in a strong gravitational field, such as that which exists near a black hole, will also create the needed time dilation. (While the wormhole that *Interstellar*'s mysterious "they" put in orbit around Saturn is not used for time travel in itself, it's not for nothing that it opens up near a black hole at the other end.).

The 1997 movie *Contact* used both the time and space travel aspects of wormholes. In this movie humans detect a signal, transmitted to them from

---

[21] Kip Thorne was also an advisor on two of the movies featuring wormholes that appear in this chapter: *Contact* and *Interstellar*.

an alien civilization, that instructs them how to build a giant machine. The machine is fitted with a capsule for one person[22]. Ellie Arroway, the discoverer of the signal, is eventually chosen to sit in the capsule when the machine is turned on. The machine transports Arroway to a distant star system via a wormhole network to meet with a representative of the aliens. She is then returned through the network to the exact point in time and space that she left.

> There was a young lady named Bright,
> Whose speed was far faster than light.
> She set out one day,
> In a relative way,
> And returned on the previous night.
>
> A. H. Reginald Buller (most likely), Botanist, 1923

Although wormhole time machines with terminologies and mechanisms similar to those described by Thorne, Morris, and Yurtsever only started popping up on screen in the years after the scientists published their paper, it is interesting to note that science fiction shows have long incorporated things reminiscent of wormholes for their time travel shenanigans, including *The Time Tunnel* (1966–1967) with its eponymous machine, and *Doctor Who* with its time vortex.

Of course, travelling back in time raises some immediate problems, such as the grandfather paradox: what happens if a time traveler were to go back in time and kill his grandfather before the conception of the time traveller's parent? Would even the tiniest interaction with the past alter the present into something unrecognizably different?[23] In general, time travel series have depicted time as extremely chaotic. A chaotic system is one that is so sensitive to its conditions, that small changes—stepping on a blade of grass, interrupting an argument, even killing one's grandfather—can have extreme consequences. This is a trope that has become so ingrained, that audiences rarely question that the situation could be any different.

Even chaotic systems can have so-called *attractors*—states that have so much "weight," that multiple starting points wind up there. Knowingly or otherwise, since its 2005 continuation, *Doctor Who* has explored this concept on several occasions, calling these attractors "fixed points." These are events that are so momentous, events that cast such a long shadow into the future, that

---

[22] In the novel, it was five people, but who wants to compete for screen time with Jodi Foster?

[23] On June 29, 2009 physicist Stephen Hawking held a party for time travelers, for which he sent out invitations after the party was over. Hawking now claims that he has experimental evidence that time travel to the past is impossible.

they are bound to happen. This concept was also explored in *Fringe* where, for example, 9/11 occurred in two parallel time streams, but the details were ever-so-slightly altered.

Figuring out the basic rules of time travel and their various implications is often a journey the audience is invited to share with the characters. Once the rules are set via the necessary exposition or story-within-a-story, stakes can be raised, tensions established, etc., without having the audience wondering why characters can't just fix any reversals they might face by hopping in the time machine.

Normally, time travel ground rules are established within the first act (if it's a movie) or pilot episode (if it's a television series), but one show that managed to turn the act of figuring out the rules into an integral part of the narrative is *Continuum* (2012–2015). *Continuum* features several characters who have arrived in present-day Vancouver from the year 2077, and in several episodes of the first season these characters directly confront the grandfather (or in *Continuum*'s case the grand*mother*) paradox when their ancestors' lives are threatened.

What exactly might be the real-world implications of time-travel is something that scientists are continuing to debate. For some—most notably, Stephen Hawking—things like the grandfather paradox are so problematic that they imply that time travel is simply impossible. This principle is known as the *chronology protection conjecture* and it means that for every theoretical time travelling mechanism plucked from the equations of Relativity (or, more recently, quantum mechanics) further investigation will inevitably reveal the existence of some flaw that rules out actually performing any such travel.

> Even if it turns out that time travel is impossible, it is important that we understand why it is impossible.
>
> Stephen Hawking, Theoretical Physicist

Some scientists believe that while it might be possible to travel back in time and interact with the past, it's impossible to change anything that has already happened—any action essentially becomes a self-fulfilling prophecy, c.f. *The Time Tunnel*, or *Babylon 5*. Other scientists have adopted variations of the *Novikov self-consistency principle*, in which time travel to the past is permissible, as long as nothing happens there that would interfere with the act of time travel itself, c.f. *Doctor Who* and its notion of "fixed points." In these scenarios if you went back and tried to kill your grandparent, something would always happen to prevent you from succeeding in your mission, although if the Novikov self-consistency principle holds, it might be possible to kill your grandparent as long as there was some other pathway, such as cloning or artificial insemination, that would lead to you being born.

Of course, the exact rules of time travel chosen by writers and producers typically depend on their immediate storytelling needs rather than a desire to join the pro- or anti-Hawking camps of theoretical physics[24]. (See the sidebar "Time Travel In The Writers' Room" for how writer/producer team Zack Stentz and Ashley Miller approached the mechanics of time travel in the 2008–2009 TV series *Terminator: The Sarah Conner Chronicles*). Generally speaking, shows or movies that are only interested in time travel as a one-shot affair adopt rules that sustain the status quo as much as possible, either adopting the self-fulfilling prophecy or strict self-consistency rules. Shows that will use time-travelling as a recurring theme tend to favor rules that allow history to be changed, not least because it opens up the possibility of dropping the cast permanently into a completely altered timeline, which can be a good way to shake up the storylines on a well-established show, as happened with both *Eureka* (2006–2012) and *Fringe*.

> If the Universe came to an end every time there was some uncertainty about what had happened in it, it would never have got beyond the first picosecond. And many of course don't. It's like a human body, you see. A few cuts and bruises here and there don't hurt it. Not even major surgery if it's done properly. Paradoxes are just the scar tissue. Time and space heal themselves up around them and people simply remember a version of events which makes as much sense as they require it to make.
>
> Douglas Adams, *Dirk Gently's Holistic Detective Agency*

For now, screenwriters and scientists are on something of a level playing field when it comes to dreaming up ways to circumvent the barriers to FTL and time travel. Part of the reason why science cannot make definitive statements is because we know that Relativity and quantum mechanics are both incomplete, and, to some extent, incompatible theories of how the universe works. Hopefully, in the next few years, scientists will discover clues—either within tools like the Large Hadron Collider, or by deploying sophisticated telescopes to probe the structure of the cosmos—that will lead them to a more complete theory. With any luck, that theory will expose cool new ideas for screenwriters looking to get their characters from A to B, or from now to then.

---

[24] For a more detailed examination of the possible rules of time travel and how and why screenwriters have used them over the years, see SAC's chapter "Visions of the Past" in *Hollywood Chemistry: When Science Met Entertainment* published by the American Chemical Society, 2013.

Hollywood Box: Time Travel in the Writers' Room

> *I don't want to talk about time travel shit. Because if we start talking about it, then we're going to be here all day talking about it, making diagrams with straws. It doesn't matter.*
>
> Old Joe (Bruce Willis), *Looper* (2012)

Zack Stentz and Ashley Miller have worked together writing and producing shows such as *Andromeda* (2000–2005) and *Fringe* (2008–2013) as well as movies such as *X-Men: First Class* (2011) and *Thor* (2011). In two separate interviews, we asked them about how they avoided getting bogged down in arguments about the nature of time travel as they developed scripts for *Terminator: The Sarah Conner Chronicles* (2008–2009).

Hollyweird Science: What's your internal working model for time travel in the Terminator universe? Does it vary depending upon storytelling needs, or is it an absolute?

STENTZ: It depends. We had hours-long discussions about time travel in the *Terminator* writers' room, which we refer to colloquially as "Going down the rabbit hole." You could literally spend hours and hours over whether this was a closed loop; whether when the characters who came back from the future altered the past; if the future that they came from continued to exist or not—you could go on and on and on about the implications… Simply to be able to tell a coherent story, at one point one writer, a very intelligent guy named John Enbom, stood up and said, "Would it be better if we just treated the future as if it was France? So when someone comes back from the future, we say 'Oh look, someone's come from France'? It's just another country that you go to and from."

That's a slightly oversimplified version of what we did, because we did deal with some of the repercussions of time travel, and whether what you do in the present affects the future and causes it to change. By and large, we liked using the "Future as France" model when doing *Terminator* stories.

MILLER: The "Future as France" was a really great insight. I agree that it's an oversimplification, I would even go as far as to say it is a *great* oversimplification. It was a way of talking to help people be a part of the conversation.

Here is why it's important: imagine that you really are telling a story about a character who comes to America from France… You're trying to break the story of what happens when this person from France is in America—the things they do, and the things they see—but for whatever reason, you get caught up in a discussion of wing dynamics, of how planes work. "What kind of plane? What sort of engine? How fast did it go? Did they sleep? How high were they? Did they pressurize the cabin?" None of that shit matters if the story you're telling is about the person who was in France dealing with being in America. So the idea of saying, "Let's just say they came from France," was a method of cutting through the parts of the discussion that weren't relevant to the drama that was happening in the "here and now."

But, by the same token, the mechanics actually *were* incredibly important, and they were incredibly important because the mechanics are governed by rules, rules are where drama comes from, and we extracted drama from those rules, from the mechanics. The two-part submarine episode that we did towards the end of the second season [2009's "Today Is the Day, Part 1 & 2"]…the drama of that relied—not entirely, but at least in large part—on the implications of the rules that emerged from the mechanics of time travel in the *Terminator* universe. There are scenes that simply could not exist if the implications of those rules weren't important.

So here's what it boils down to: the best teachers are people who can take incredibly difficult, convoluted, complicated concepts and communicate their implications in an elegant way. Saying "The Future is France," saying that somebody who comes to us from the future are simply coming from another country, is an elegant way of describing the consequences of something that is, otherwise, incredibly complicated.

10
Moving in Stereo: Parallel Universes

I was good at math and science, and I got lots of degrees in lots of things, but in a parallel universe, I probably became a chef.
Nathan Myhrvold, Former CTO of Microsoft, *Author of Modernist Cuisine*

The most watched programme on the BBC, after the news, is probably Doctor Who. *What has happened is that science fiction has been subsumed into modern literature. There are grandparents out there who speak Klingon, who are quite capable of holding down a job. No one would think twice now about a parallel universe.*
Sir Terry Pratchett, Author

The Universe forces those who live in it to understand it. Those creatures who find everyday experience a muddled jumble of events with no predictability, no regularity, are in grave peril. The Universe belongs to those who, at least to some degree, have figured it out.
Carl Sagan, Planetary Scientist

The notion that there are multiple self-contained universes, or alternate realities, coexisting simultaneously with our own, dates back centuries. The terms "infinite worlds," "multiple universes," "countless universes," and "infinite universes," can be found in Hindu Puranic literature, Buddhist texts, even in Kabbalah. For that matter, Christian Heaven and Hell can be considered different planes of existence[1] that coexist simultaneously with ours.

Even within literary and cinematic fantasy, there are no shortage of examples—*Narnia, Alice in Wonderland, The Nightmare Before Christmas* (1993), *Coraline* (2009), even the *Harry Potter* (2001–2011) movies—where characters pass through some sort of interface (e.g., through a door, down a hole, into the fireplace, or under a creepy archway) into a neighboring realm where the physical laws may or may not reflect those of our daily experience.

---

[1] The Robin Williams film *What Dreams May Come* (1998) took place almost entirely on a different plane of existence than our own.

In science the definition of "universe" has not only changed over time, but the definition depends upon the discipline of the scientist using the term. Three centuries ago, the "universe" was everything—all that there is. That was when the stars of the Milky Way were thought to be the sum total of the Universe, though. When astronomer William Herschel performed a more detailed survey of the fuzzy patches of light in the sky, of many of the same objects surveyed earlier by Messier but at greater magnification, he observed that some revealed a spiral structure. It was philosopher Immanuel Kant who called these *island universes* (recall the great distance to these objects was confirmed by Edwin Hubble in the 1920s). Since the "universe" was intended to reflect all there is, "island universes" quickly fell out of favor, to be replaced by "galaxy."

Nowadays, *universe* has come to mean "the sphere of influence encircled by the effects of the Big Bang 13.8 billion years ago" to astronomers, but to theoretical physicists, *universe* has a different meaning, even potentially multiple meanings. Discoveries in theoretical physics suggest that there might exist other universes, parallel to ours, that we can not observe through a telescope. There might even be multitudes of alternate realities spawned from our own every moment. With its scope as "all there is" superseded, new terms came into being to describe the collection of all potential universes—the most common of which is *multiverse*.

> There are many of us thinking of one version of parallel universe theory or another. If it's all a lot of nonsense, then it's a lot of wasted effort going into this far-out idea. But if this idea is correct, it is a fantastic upheaval in our understanding.
>
> Brian Greene, Theoretical Physicist, Author, *The Elegant Universe*

Two prominent theoretical physicists—Max Tegmark from MIT and Brian Greene from Columbia—have categorized the different types of parallel universes that could theoretically exist, with Greene's nine models being a superset of Tegmark's four (and a superset of the types of universe that have been portrayed on both big and little screens).

In some of the universes that these physicists describe either the physical laws or fundamental physical constants are different than those in our Universe. While interesting from an intellectual standpoint, universes where the physical laws vary significantly from ours are, in general, dramatically uninteresting in a fictional narrative: a universe that is so fundamentally alien that an audience can barely comprehend it, if at all, hardly wins high the "relatibility" category.

If there are parallel universes, would we ever be able to detect them? Travel to them? For some multiverse models, the surprising answer is, "Yes. To both."

10.1 The Simulated Universe

> Is all that we see or seem
> But a dream within a dream?
>
> Edgar Allan Poe, *A Dream within a Dream*

The Simulated Universe, or Simulation Hypothesis, has high-tech trapping but actually dates back the philosophical musings of the ancient world. Circa 500 BC philosophers like Parmenides of Elea, Zeno of Elea, and Plato questioned whether what humans perceive as reality is some form of illusion or dream.

In the 1960s and 1970s, declaring a sequence of events that the audience had previously taken for reality as just "all a dream" was a too-commonly-used trope that saved many a screenwriter who had painted himself into a corner. This trope reached its highest expression when the entire ninth season of the long-running series *Dallas* turned out to be an extended dream of Pamela Barnes Ewing's at the opening of the tenth season in 1986. This effectively killed any subsequent semi-serious attempt to use the trope again, although a variant is still used to portray the inner thoughts of characters, playing out their fears or hopes for a few moments on screen before reality reasserts itself.

The Simulation Hypothesis is the modern technological manifestation of "it's all a dream." In 2003, Oxford University philosopher Nick Bostrom published a paper[2] that suggested that unbeknownst to us all that we are, and the reality around us, might be a computer simulation like *Sim City*.

Bostrom's rationale was threefold. He says that soon the state of computation will be such that humans (or post-humans) could simulate human thought processes on a computer. Assume, also, that these simulated minds are conscious (some modern philosophers like John Searle would violently disagree at this point, but we'll let Bostrom's claim stand for now). It would be simple then for many sets of these simulations to be run simultaneously. Multiple universes, then, would simply be different simulations running on different CPUs with different (or not) initial conditions.

Were that the case, the number of simulated consciousness would rapidly outnumber the biological. Once *somebody* reaches that level of technological sophistication, by simple statistics you are more likely to be a simulated consciousness than natural. (Being the original nonsimulated consciousness is like being a lottery winner. Someone has to be a winner, but the chances are slim that it's you.) The scenario that our consciousness is simulated on a com-

---

[2] Bostrom N (2003) Are you living in a computer simulation? *Philosophical Quarterly* 57(211): 243–255 http://www.simulation-argument.com.

puter, and reality simply a hyperrealistic virtual reality simulation, has been explored in films like the *Matrix* series (1999–2003) and *Total Recall* (1990, 2012), as well as the series *Star Trek: The Next Generation* (1987–1994), and *Eureka* (2006–2012).

Following up on Bostrom's paper, in 2007 Cambridge University professor of mathematical sciences John D. Barrow asked, "What we might expect to see if we made scientific observations from within a simulated reality?"[3] Barrow suggested that an imperfect virtual reality simulation would contain observable glitches or errors. When Pixar simulates a scene in a story, complete with reflections from water, mirrors, or windows, they don't start with Maxwell's equations of electromagnetism, but make simplifications along the way to save on memory and processing requirements. Barrow's approach takes it that for the overwhelming majority of the time, these simplifications are good enough, but ever so often, there might be observable momentary glitches. There might also be smaller errors, unnoticeable at first, that accumulate over time to the point where they do create observable errors. Simulation errors might require periodic software patches, which might manifest as discontinuities in reality (a la when the rebels in *The Matrix* spot the same cat walking past them in the same way twice). Much of Barrow's writing was the basis of the 2012 *Eureka* episode "Force Quit."

> The universe is big, it's vast, and complicated, and ridiculous. And sometimes, very rarely, impossible things just happen and we call them miracles.
>
> The Doctor, *Doctor Who*, "The Pandorica Opens"

In 2012, the simulation model gained renewed attention when physicists indentified specific physical observations that could determine if we exist within a simulation (that is, they acknowledged, if we were not pre-programmed never to discover the true nature of our existence). In a natural universe, the expected distribution of cosmic rays—high-energy subatomic particles radiated from some astronomical phenomena—would be isotropic, or equally-likely to come from any direction. If we lived within a simulation that was gridded, i.e., positions were quantized or pixilated, then cosmic rays would be strongly biased along orthogonal (perpendicular) directions.

The argument is not without its logical flaws, but it does incite interesting lunch-time conversations. The remaining parallel universe models assume that we, as well as our Universe and (potentially) others, are naturally-occurring rather than simulated.

---

[3] Barrow JD (2007) Living in a Simulated Universe. In: Carr B (ed) *Universe or Multiverse?* Cambridge University Press, Cambridge/New York, pp. 481–486.

10.2 The Multiple Big Bang or Quilted Model

Our *Hubble volume* or *Hubble sphere* is often mistaken for the size of our Universe—it a sphere of radius 13.8 billion light years in radius (about 1.1×10^{31} cubic light years), which is the speed of light propagated over the time since the Big Bang, and ever-expanding at the speed of light. Given the "stretchiness" of spacetime, our actual universe is much larger, but our Hubble volume is the part of that we can observe[4], and the boundary of the observable Universe is called the *cosmic horizon*.

What if the "all that there is" is much larger than our Hubble volume, and recent observations suggest that it is, and there were multiple *local* Big Bangs beyond our cosmic horizon? This would create many expansions of spacetime and matter—many universes—with the same physical laws as ours, none of which could see one another. This is the basis for Greene's *Quilted Universe*, or Tegmark's *Level-I Universe* (Fig. 10.1).

If all of these local Big Bangs are immersed within an infinite volume, and if the matter in each universe is distributed in, more or less, the same way as the others, then there is only a finite number of ways matter can be arranged. In that event, every possible arrangement of matter, every event, every you, will be repeated an infinite number of times. It is the speed of light, and the sheer distance of these other universes, that prevent us from being aware of our other "selves."

Hollywood has done stories on alternate Earths of this type. In "Miri," a 1966 episode of *Star Trek*, the crew of the *Enterprise* discover a planet whose geography is the same as Earth. The only inhabitants of the planet are children, though, as everybody expresses a lethal disease after entering puberty. In the 1969 film *Journey to the Far Side of the Sun*[5], after a robotic probe discovers a planet that shares Earth's orbit, but on the opposite side of the Sun, the European Space Exploration Council sends two astronauts on a mission to the planet, only to discover that it is a mirror image of Earth in more than one way.

In these dramas, though, the lines are very blurry between the narratives being descriptions of parallel Earths and Earths that arose from *convergent evolution*. Convergent evolution, a term from biology, occurs when similar types of animals in disconnected ecosystems evolve similar traits and fill similar ecological niches (think how similar birds and bats are, yet one descends from dinosaurs and the other is basically a flying rodent). It is highly unlikely that a parallel Earth, or a parallel you, in this model exists within this galaxy, or in

---

[4] The Hubble volume is actually slightly larger than the observable Universe because the expansion of the Universe has accelerated. At this point in time, though, the two values are qualitatively close.

[5] Known as *Doppelgänger* outside of the U.S., it was written by Gerry and Sylvia Anderson who brought you the series *Thunderbirds*, *UFO*, and *Space: 1999*.

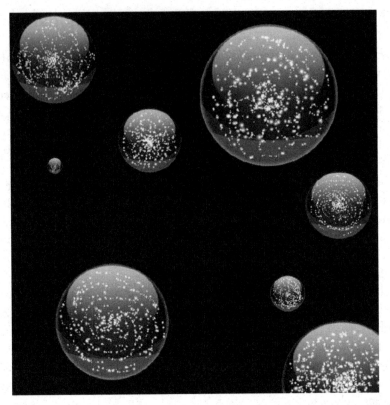

Fig. 10.1 A level-I multiverse with multiple hubble volumes. (Illustration by Chelsii Carter)

any galaxy we can see in a telescope. In a real quilted universe, Max Tegmark writes[6], "A crude estimate suggests that the closest identical copy of you is about $10^{10^{29}}$ m away. About $10^{10^{91}}$ m away, there should be a sphere of radius 100 light-years identical to the one centered here, so all perceptions that we have during the next century will be identical to those of our counterparts over there. About $10^{101^{15}}$ m away, there should be an entire Hubble volume identical to ours."

10.3 Braaaaaanes

The difference between the next model and the quilted model is that universes in the quilted model are all three-dimensional volumes (four, if you include time) within a larger three-dimensional volume. The universe for any observer

---

[6] Tegmark M (2003) Parallel Universes. *Scientific American*.

is the Hubble volume that is defined by the limit of expansion of mass after a local Big Bang. In theory, if you could travel fast enough, and outpace the "stretchiness" of the space of your local universe, you could travel to another Hubble volume. In the *brane* (short for membrane) multiverse model, on the other hand, each universe is one of many coexisting four-dimensional volumes immersed within a higher-dimensional space called the *bulk*. Each brane is disconnected from the others, and matter is bound to its brane[7]. While universes in the quilted model are like blobs of wax floating within a lava lamp, brane universes are like drawings on different pages of a book—you can't go from one picture to another, although in the science fiction comedy web series Other Space (2015–) a "tear" between membranes strands a luckless spaceship crew in another universe.

Surprisingly, physicists are devising experiments to determine if this model reflects reality. In an interview with NPR[8], Greene says:

> If we are living on one of these giant membranes, then the following can happen: When you slam particles together—which is what happens at the LHC—some debris from those collisions can be ejected off of our membrane and be ejected into the greater cosmos in which our membrane floats. If that happens, that debris will take away some energy. So if we measure the amount of energy just before the protons collide and compare it with the amount of energy just after they collide, if there's a little less after—and it's less in just the right way—it would indicate that some had flown off, indicating that this membrane picture is correct.

Not only is there is no guarantee that the physical laws that prevail in our universe are the same as those in the bulk, there is no guarantee that the physical laws of our Universe rule in other universes (which is the thinking behind Tegmark's Level-II Universes and some of Greene's parallel universe models we will overlook here). There is certainly no guarantee that there is another "you" (or a multitude of "yous") existing within separate branes in this model—though, if the universes were old enough, and matter distributed evenly enough, by the same argument for an alternate "you" in the quilted universe, there is very likely another "you" on another brane (Fig. 10.2).

Popular in 1960s' and 1970s' science fiction was a concept called the *Fifth Dimension*. In some instances, the concept referred simply to an unexplored or strange parallel realm. It others it is an extra dimensionality to the universe

---

[7] Though gravity might not be. Some physicists are entertaining the notion that gravity "leaks" off of the brane of its origin to the bulk or nearby branes. This might explain why gravity is so much weaker than the other three fundamental forces, i.e., why a cheap fridge magnet can hold up a piece of paper against the gravitational force generated by an entire planet.

[8] http://www.npr.org/2011/01/24/132932268/a-physicist-explains-why-parallel-universes-may-exist.

Fig. 10.2 A level-II multiverse with parallel branes immersed in a larger bulk like the sheets of a book. (Illustration by Chelsii Carter)

necessary to travel through time—the logic being that if it takes a fourth dimension, time, to travel in space, then it must imply a fifth for motion through time. One might also argue that the fifth dimension is the bulk, the higher-dimensional space in which our spacetime is immersed. In the 2009 alternate reality episode of *Eureka* entitled "Insane in the p-Brane," the term "fifth dimension" is used synonymously as the "theoretical membrane barrier between dimensions"; in other words, the bulk.

> There is a fifth dimension, beyond that which is known to man. It is a dimension as vast as space and as timeless as infinity. It is the middle ground between light and shadow, between science and superstition, and it lies between the pit of man's fears and the summit of his knowledge. This is the dimension of imagination. It is an area which we call The Twilight Zone.
>
> Rod Serling, Creator and Host, *The Twilight Zone*

From a dramatic standpoint, assuming identical physical laws, parallel universes in the brane model would look, behave, and evolve in much the same manner as in the quilted model. The next model is defined by the manner in which events diverge in different universes.

Fig. 10.3 Allison Blake (Salli Richardson), Sheriff Carter (Colin Ferguson), Henry Deacon (Joe Morton), and Kim Anderson (Tamilyn Tomita) face the chilling truth of an alternate reality in the 2006 Eureka episode "Once in a Lifetime." (Copyright © NBC Universal Television, Universal Cable Productions. Image credit: Photofest)

10.3.1 Quantum or "Many Worlds" Multiverse

Although the terms *alternate reality* and *parallel universe* are often used interchangeably, the terms are overlapping, but not identical. The quantum multiverse model—Greene's Quantum Multiverse or Tegmark's Level-III Universe—is the parallel universe model where the terms are most closely synonymous. The term alternate reality implies that a parallel universe and ours are variations on a theme—that there is a relationship between them, and that events leading up to the present share an identical origin and evolution up until a point where one universe bifurcated into two. The bifurcation happens each time a quantum event occurs (Fig. 10.3).

The quantum multiverse is an outcome of the *Many Worlds* interpretation of quantum mechanics, proposed by Hugh Everett in 1957. The many-worlds hypothesis is an attempt to solve a very deep problem in quantum mechanics, the Schrödinger's Cat problem. Recall that for the Schrödinger's Cat *gedanken* experiment, imagine a cat sealed in a box, with a 50–50 chance of living or dying in a given time period T. After time T elapses, what can we say about the cat? Nothing. Until we observe it, it exists in an intermediate, indeterminable, state.

Everett's many-worlds solution was to say that the idea of the universe somehow choosing between an alive cat and a dead cat was wrong. Instead,

Fig. 10.4 Mirror universe spock figures out that Captain Kirk is not his Captain Kirk in the 1967 *Star Trek* episode "Mirror Mirror." (Copyright © Paramount Television. Image credit: Photofest)

each quantum event causes the universe to bifurcate, spawning multiple universes—in the case of Schrödinger's Cat, a universe where the cat is alive, and a universe where the cat is dead. From that point, the timelines of each universe diverge.

> If a coin comes down heads, that means that the possibility of its coming down tails has collapsed. Until that moment the two possibilities were equal. But on another world, it does come down tails. And when that happens, the two worlds split apart.
>
> Philip Pullman, *The Golden Compass*

Alternate reality parallel universes are common in science fiction—often these plots are used just as a way to have some fun with established characters, such as in the 1967 "Mirror, Mirror" episode of the original *Star Trek* series (1966–1969). A transporter malfunction pushes Kirk, Scotty, Uhura, and McCoy into an alternate universe where the *Enterprise* is a warship of the Terran Empire (Fig. 10.4). (The costume choice to distinguish the alternate Spock in this episode by giving him a goatee is responsible for one the most enduring visual tropes for *any* show with a doppelganger: goatees = evil.) Seeing Kirk as a tyrant and the rest of the crew as venial, backstabbing versions of themselves was a welcome break from the usual straight-laced decorum Federation starship crews were expected to maintain. Although the Mirror universe would be revisited a few times during the run of 1993–1999s *Deep*

Space Nine, and 2001–2005s *Enterprise*, it never had an enduring impact on the narrative course of any show or movie in the *Star Trek* franchise.

While it might initially seem counterintuitive, what multiverse exists, or the multiverse in which a writer sets a story, has important time travel implications. We discussed in Chapter Nine how time travel is frequently depicted as chaotic: go back in time, sneeze at the wrong time, and humanity first lands on the Moon in 1985. One way this extreme sensitivity might readily be explained, even expected, is that if the quantum universe model is correct, and, at each moment, multitudes of realities come into existence as a result of "decisions" at the quantum level.

If there is a quantum multiverse, then as time travelers journey backwards in time, the ever-branching tree of quantum probabilities would, instead, merge like tributaries of a river. Moving forward again, not only would it be reasonable that you might not take the same path forward, it could be argued that it is *impossible* for you to take the same path since the introduction of you, traveling companions, and whatever time machine you employ, have altered this reality. So changing the past simply creates a new timeline without necessarily wiping out the original timeline from which you, and any fellow travelers, originated.

The Many Worlds model also renders some paradoxes nonparadoxical. Travel backwards in time, kill your grandfather, you continue on in your current branch of the Many Worlds tree, while the timeline that give rise to your birth remains unaltered. At the same time, this can complicate the "past gives rise to the future gives rise to the past" sort of stories that are common on *Doctor Who*, featured in films like *Frequency* (2000) and the *Terminator* films, and averted in *Looper* (2012).

The timeline aspect in shows featuring parallel universes is often left in the background—characters might compare notes to find the point of divergence between universes, as in the 1999 "Point of View" episode of *Stargate: SG-1* (1997–2007)—but there is nothing to be done about the divergence itself. However, dealing with divergences has become a key element of the series *Continuum* (2012–) and also played a role in the *Primeval* franchise (2007–2013).

Is Everett's vision of endlessly branching worlds true? At the moment we have no way of saying, although if those parallel universes *are* out there, we don't really have an inkling of how we might visit them.

> The next time you check your moves in the mirror and reflect on how special you are, consider that somewhere in this universe or in another parallel universe, your double might be doing the same. This would be the ultimate Copernican Revolution. Not only are we not special, we could be infinitely ordinary.
>
> Seth Shostak, Astronomer, SETI Institute

10.4 Hybrid Parallel Universes

If there are universes existing on other branes, or other Hubble volumes, then although they may or may not be versions of you inhabiting each of them, there is nothing precluding other forms of life. There is also nothing precluding a quantum multiverse existing and evolving within each brane or Hubble volume. That the multiverse may be a hybrid of all possible parallel universes is the basis for Tegmark's Level-IV Universe and Greene's Ultimate Ensemble.

In the quilted and brane universe models, traveling from one universe to another may involve traveling many times faster than light, or "tunneling" through some interstitial realm like the bulk (cf the Dalek Void Ship in the 2006 "Army of Ghosts" episode of *Doctor Who*). How does one travel from one reality to another in a quantum universe, when what lies between "here" and "there" is not a vast distance or a physical barrier, but the outcome of a quantum event in the past? The best science fiction answer to this would be to travel backwards in time—in that instance timelines converge rather than diverge, then move forward again along a different quantum path. Choosing which path, though, would be highly problematic.

In the novel, and 2003 film, *Timeline*, Michael Crichton reversed these notions. The novel was based upon the premise that there might be nearby branes where history mirrors our own, but with an offset in time. Researchers used this device to travel to 14th-century France. So rather than using travel backwards in time to visit alternate quantum universes, Crichton relied upon travel to a hybrid model universe to implement what amounted to time travel.

Parallel universes depicted in fictional narratives are most often portrayed as a hybrid between either a quilted universe or a brane universe and a quantum universe, and several television series have made substantive attempts to grapple with the notion of parallel universes of this nature. *Sliders* (1995–2000) featured its cast dropping into a different alternate universe nearly every week. On *Fringe* (2008–2013) the existence of a parallel universe became a defining element of the show, especially after the third season. Rather than just being played for comic effect, the alternate versions of the core characters of Olivia Dunham and Walter Bishop—dubbed Fauxlivia and Walternate—became almost as prominent and well-characterized as those from this Universe, as the consequences of connecting two divergent realities played out on scales ranging from the personal to the cosmic.

> The mystery of life isn't a problem to solve, but a reality to experience.
>
> Frank Herbert *Dune*

11
Braver Newer Worlds

You cannot look up at the night sky on the Planet Earth and not wonder what it's like to be up there amongst the stars.
Tom Hanks, *Actor*

Here men from the planet Earth first set foot upon the Moon. July 1969 AD. We came in peace for all mankind.
Neil Armstrong, Astronaut, *Apollo 11*

In terms of Defiance, *this world doesn't actually exist. My job is to convince our audience that it* could *exist.*
Kevin Murphy, Showrunner, *Defiance*

A young star, its fusion furnace recently ignited, is surrounded by a disk of gas and dust, much like Saturn is encircled by its paper-thin rings. The disk is composed mostly of hydrogen, the most basic element and the building block from which all other elements were constructed, but there are other gases like helium.

The disk starts off hot, but as it cools, compounds condense out of the protoplanetary fog. Close to the star, where it is hot, silicate dusts—similar, in many respects, to the dust you might wipe off a coffee table—and metals like iron and nickel form tiny orbiting grains. Farther out, beyond a point known as the *frost line*, tiny ice crystals form in addition to the silicates and metals: water ice, carbon dioxide, ammonia, and methane.

On rare occasion, grains collide. On even rarer occasion, they stick together. Over time, grains collide to form ever-larger grains. Eventually, some grains become the size of boulders, some grow to the size of mountains. These larger chunks of solid material are called *planetesimals*. Planetesimals, still immersed within a gassy disk, collide with other planetesimals to grow still larger, until some grow to sizes large enough where they have a nontrivial amount of gravitational attraction.

Capitalistic growth takes over, where the rich get richer. Objects with the most mass gravitationally attract more material, giving them more mass, giving them a stronger gravitational attraction, giving them the ability to attract

Fig. 11.1 The evolution of Earth. (Photo credit: European Southern Observatory)

more mass, and so on. Planetesimals that are attracted towards these ever-growing planetary embryos, but that somehow manage to miss impacting, are slung both towards the hot young sun, and into the distant icy reaches of space, some never to return. This is a very chaotic time, and huge impacts are common as planetary embryos accrete millions of planetesimals to grow into young planets (Fig. 11.1).

Almost as if it is sensitive to, even empathetic with, this upheaval, the Sun enters a tumultuous phase as well, where it becomes more active and prone to outbursts, like a pubescent teenager. Strong stellar winds now blow from the star—streams of plasma and subatomic particles—that blow the remaining gas of the inner system outwards. The inner planets are composed of rocks and metals, and are too small to grab onto much gas as it is pushed past. The outer planet cores, composed of some metal, a little rock, and a lot of ice, are large enough to hold onto the gas first-come, first-served.

The final remaining planetesimals are either incorporated into the planets, ejected from the system, or form a rubble of asteroids, mostly concentrated in the gravitationally perturbed no-mans land between Mars and Jupiter. The final product is a young stable star, inner planets composed of rock and metal, called *terrestrial planets*, and outer planets with large cores and even larger gassy envelopes, called *jovian planets*.

Thus the Solar System is born[1].

We have examined the basic stuff of the Universe, and the rules that govern it. We've explored the Universe itself: its birth, likely fates, and whether or not it has siblings. We've explored how one might engineer means to ply the vastness of the cosmos within a human lifetime. From a storytelling standpoint, all of that is interesting, it is enabling, it is backstory, but it is difficult to boldy go, if you have nowhere to boldly go *to*. Science understands fairly well how the planets of the Solar System formed, but—as telescopic data about exoplanets is making clear—different systems around different suns may have evolved in very alien ways.

One of the strength's of screen science fiction is its ability to put us right there on the surface of an alien world. Think of the icy expanses of Hoth. The lush jungles of Pandora. The endless deserts of Arrakis. These are worlds so vividly created that they have, to a degree, transcended the movies that introduced them to us—namely *The Empire Strikes Back* (1980), *Avatar* (2009), and *Dune* (1984). Whether sketched out in a few scenes as with Hoth, or in full 3D IMAX glory like Pandora, a sense of place is transmitted that makes these worlds feel more than just another backdrop. Somehow we feel like *we've* crunched across Hoth's snowpack, pushed through Pandora's dense foliage, or watched the air shimmering above the scorching sands of Arrakis.

Some fictional worlds act as a character in their own right. Recall the many episodes in the *Star Trek* franchise in which the central conflict is crew vs. planet, such as the 1967 story "This Side of Paradise" (local plant spores alter the personalities of the crew of the *Enterprise*), or 1996's "The Ascent"

---

[1] One of the most popular episodes of the reimagined *Battlestar Galactica*, "Scar," was set in one such protoplanetary disk. That explains why the environment between tumbling rocks appeared hazy.

Fig. 11.2 Humans share a terraformed Earth with seven alien races in SyFy's *Defiance*. Pictured are Castithans Datak Tarr (*Tony Curran*) and Stahma Tarr (*Jaime Murray*). (Copyright © Universal Cable Productions. Image credit: MovieStillsDB.com)

(stranded inhabitants of Deep Space Nine must climb a mountain to transmit a distress signal). The terraformed Earth of 2046 is as much a character in *Defiance* (2012–) as Nolan, Irisa, or any of the Tarrs or McCawleys.

Creating a believable world on screen is a highly collaborative affair. A good writer will understand the limits of what can and can't be done given the amount of time and money available when writing stage directions, but turning those directions into sets and special effects relies heavily on the skills of directors, location scouts, set designers and—for the productions that can afford them—visual effects artists.

Kevin Murphy, showrunner for SyFy's *Defiance*, describes the lengths to which his team goes to create the tapestry of the show set on a fractured and newly alien Earth, "We're building a history, we're doing documents on mythology, we're doing the science, we're dealing with alien culture, we need to know the shape of the original Votan solar system, we need to know that they have dual suns, and how all that works. But at the same time we're also developing the Castithan caste system, or liro. Anywhere I can find somebody that's got life experience to add to this big tapestry of the world, I'm trying to take advantage of it" (Fig. 11.2).

North American shows that tend towards a "planet of the week" theme, such as the *Star Trek* or *Stargate* franchises, typically use broad brushes to present easily-understood planets that look a lot like places within a day's drive of Los Angeles (scrubby desert), Vancouver (misty forests), or Toronto (not-quite-as-misty forests). *Stargate SG-1* humorously hung a lantern on this whole business when team leader Jack O'Neill complained about the similar environments of the worlds they visited in the 1999 episode "Demons", quipping "Trees, trees, and more trees. What a wonderful green universe we live in." On the other side of the Atlantic, episodes of *Doctor Who* (1963–) and *Blake's 7* (1978–1981) typically wind up being set on planets that share much in common with Welsh rock quarries.

Productions that focus on a single planet, such as the movie *Dune* or TV series like *Earth 2* (1994–1995) and *Defiance* will typically take the time to build up a collection of disparate locations and alien flora and fauna, even within the confines of a limited number of filming locations. Perhaps none have done so as successfully as *Avatar*, which depicted a tumultuous ecosystem that is the very embodiment of Tennyson's phrase, "Nature, red in tooth and claw." Indeed, the ecosystem—which supports a global neural network, carnivorous plants, and an indigenous sentient population with a "genuine" alien language created by linguist Paul Frommer—is considerably more complex and original in conception and execution than the film's actual plot, which is a predictable collection of movie tropes[2].

Even when a single planet happens to be the focus of the action, screenplays that incorporate faster-than-light travel tend to establish that world as just one of a vast armada of planets floating about the galaxy. In this depiction of planetary abundance, generations of writers took something of a gamble, albeit out of dramatic necessity—characters need somewhere to explore/save/hangout other than their spaceships. In reality, it wasn't until 1992 that astronomers made the first definitive observation of a single planet outside the Solar System.

The writers' gamble has paid off. In the last ten years or so, thanks to improved detection methods, there has been an explosion in the number of confirmed detections of planets orbiting other stars, called extrasolar planets, or exoplanets: over 1800 at the time of writing, with many more candidate systems awaiting confirmation. Admittedly, perhaps it was not that much of a gamble: there are roughly 1/10th of a mole[3] of stars in the Universe, so even if one in a trillion stars had planets, it would mean that the observable Universe is populated with 10 billion stars with planets—more than one for every person on Earth. Still, there is a big difference between finding a planet, and finding one that we know for sure can support life, particularly intelligent life—a search that continues to this day.

---

[2] Which is why *Avatar* is frequently referred to as *Dances with Na'vi*.

[3] A *mole* is a term borrowed from chemistry, also known as Avogadro's Number, it is 6.02×10^{23}. There is one mole of carbon atoms in a 12-gram lump of carbon.

11.1 Location, Location, Location: The Right Kind of Star

In Chapter Eight we introduced the Hertzsprung-Russel diagram, illuminating how astronomers classify stars based upon their spectral type. Science fiction has been efficient at propagating two myths: that our star, Sol, is "just an average star," and that any star can support planets harboring life.

In the history of astronomy, there have been many humanocentric beliefs that have fallen by the wayside thanks to better and better observation. We know now that Earth is not the center of the Universe, the Galaxy, or even the Solar System. Although Sol may be rare, it is not *that* rare. There are eight planets in the Solar System, and likely billions upon billions orbiting Sol-like stars in the Universe. There is one humanocentric argument, though, that persists, and that is the belief that if we find life "out there," in particular intelligent life, it will be from a planet orbiting one of these cousins of our own Sol.

What really are the limits on where we might find life? Scientists who study life and, in particular, the prospect of finding it somewhere other than Earth, believe that there is a good argument to be made for water being a necessary component for life. Water is an excellent solvent, and the temperature range where it is a liquid encompasses temperatures where some interesting and complex chemistry can occur. Even if it is not an absolute requirement, the presence of water helps narrow the search to life that we know how to identify[4]. Surrounding each star is a zone, known as the *habitable zone*, in which liquid water could exist on a planet—it's where the temperature is not too hot, not too cold, it's just right. For obvious reasons, the habitable zone is, more commonly, called the *Goldilocks Zone*. The bigger the star, the farther away and broader this zone is (Fig. 11.3).

For stars larger than Sol, the Goldilocks zone is larger than that for the Solar System. Still, we would be very unlikely to find life on planets orbiting stars much larger than our own simply because the bigger the star, the shorter the lifetime. It took 800 million years after the formation of the Solar System for the first single-celled life to form on Earth. Even though both planets and life may form faster in other systems, they can't form *that* much faster. Stars like Rigel live their entire lives between 8 and 80 times over in the span of time it took life to form on Earth.

In the *Star Trek* universe, Rigel (the left knee of the constellation Orion the hunter), is one of the most populated star systems, boasting several inhabited

---

[4] The original series *Star Trek* episode "The Devil in the Dark" explored a situation where, even in the 23rd century, humans colonizing new planets had difficulty identifying truly unusual lifeforms.

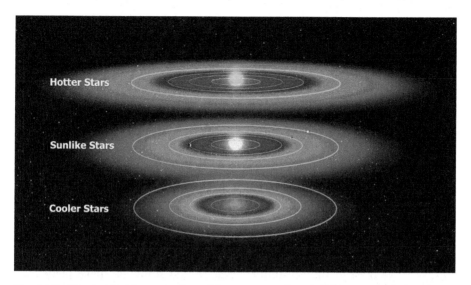

Fig. 11.3 The habitable zones for a *blue* star, a *yellow* Sol-like star, and a small *red* star. The hotter the star, the farther away, and wider, the habitable zones. In each instance, the *green orbit* indicates the orbital distance where a planet would receive the same amount of sunlight as Earth. (Credit: NASA/Kepler Mission/Dana Berry)

worlds that were visited often over five series and eleven movies, and mentioned many more times. Unfortunately, Rigel is a big, bright blue star, and one of the last star systems that would ever have planets, in particular habitable ones. Not only will Rigel live a short lifetime, it has already begun the transition to its death phase which will one day culminate in a supernova. A star in a similar state is Deneb. Sorry, but say "Goodbye" to *Trek*'s Denebian Slime Devil as well.

Yet another star that both has a short lifetime, and is in the early stages of its death march, is Canopus: home of Frank Herbert's mythical world Arrakis, also known as Dune. Canopus simply lives too short a lifetime to have habitable planets, and the only way Arrakis could be inhabited would be if it had been terraformed, and the only way the Sahi-hulud (its legendary sandworms) could exist is if they had evolved on a different planet, and were then introduced into the Arrakean environment[5] (Fig. 11.4).

At the other end of the size spectrum, small stars are believed to have incredibly long lifespans, so time is not an issue. Other than by interactions with other stars, there may have not yet been a red dwarf, or M-class, star end its life by "natural causes" in the entire history of the Universe. The reason why these stars are also unlikely candidates for life is a little more complicated.

---

[5] Contrarily, implied in the books is that Arrakis was *more* Earthlike before the introduction of the Shai-hulud, not less.

Fig. 11.4 Fremen ride the gigantic Shai-hulud (*sandworms*), of the planet Arrakis in *Dune* (1984). (Copyright © Universal Pictures. Image credit: MovieStillsDB.com)

Smaller stars have much smaller Goldilocks zones in which a life-sustaining planet could form (Fig. 11.3). With so many stars, and so many of them small red stars, the overwhelming numbers dictate that some will have planets in the zone. One such planet is Gliese 581, about 20 light-years away in the constellation Libra. Three planets—a, b, and c—were discovered orbiting Gliese 581, and Gliese 581c orbits at the inner edge of the system's habitable zone (Fig. 11.5). Gliese 581c is almost certainly an extrasolar Venus, rather than Earth-like (more on that later). Scientists then announced finding planets d, e, f, and g. Not only was d at the outer edge of the habitable zone, but g was smack-dab in the middle. Three planets in the habitable zone of a nearby star! How exciting! It turns out that although Gliese 581e may exist, planets d, f, and g probably do not[6]—their detection is likely an observational artifact from a very active star.

This brings us to why smaller stars are unlikely to have habitable planets—they are very prone to highly active periods, and eruptions of stellar flares. Stellar flares are bursts of starlight and subatomic particles, similar to a massive outpouring of solar wind, that could sterilize any biological activity on a young planet—especially one without the protective envelope of a magnetic field like Earth's. Since the habitable zones of small stars are narrow and close to the star, any planets in the zone would receive a much higher dose of harmful radiation than does Earth from a solar flare.

Similar to Gliese 581, the star Gliese 667 C is the smallest member of a trinary star system, 22 light-years away in the constellation Scorpius (though still larger than Gliese 581). The two larger members of the system, stars A and B, pass very close to one another in their orbits, and likely do not have habitable planets. Gliese 667 C orbits distantly, and appears to have several

---

[6] Robertson P et al. (2014) Stellar activity masquerading as planets in the habitable zone of the M dwarf Gliese 581. *Science* 345(6195): 440–445.

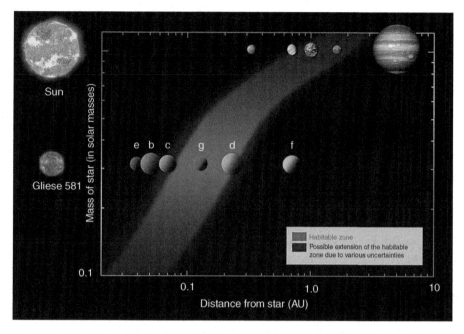

Fig. 11.5 The Gliese 581 System. The *blue* swath represents the habitable zone as a function of stellar size. The axes are logarithmic, with powers of 10 displayed at equal intervals. This means that although the habitable zone of Gliese 581 appears wider than Sol's, Sol's is actually about five times as wide. Note, also, that the existence of planets d, f, and g is questionable. (Photo Credit: European Southern Observatory)

planets, some of which are in the habitable zone. Gliese 667 C is also what is known as a flare star, a star that undergoes brief, but dramatic periods of increased brightness, likely associated with flare activity. This lessens the likelihood that planets orbiting within its Goldilocks zone are habitable. Other famous flare stars are Proxima Centauri, Wolf 359[7], and Barnard's Star.

11.2 Location, Location, Location II: The Right Kind of Galaxy

Not only will life likely be limited to planets orbiting certain classes of stars only, life is also most likely to be found in certain types of galaxies only. We have already visited Hubble's "tuning fork" diagram, and saw that galaxies come in four basic varieties: ellipticals, spirals, barred spirals, and irregulars. The two flavors of spiral galaxies are far more likely to support stars with planets, and planets with life, than ellipticals or irregulars.

---

[7] Famous for the Battle of Wolf 359 from the *Star Trek* franchise.

Fig. 11.6 Elliptical galaxy M87 (*left*) and spiral galaxy NGC 5033 (*right*). (Image credit: M87: StSci; NGC 5033 Wikimedia Commons taken by Joseph D. Schulman)

This is because elements like carbon, nitrogen, and oxygen, essential for life as we know it, were fused in the intense heat and pressure within stellar cores. More massive elements, crucial for the construction of terrestrial planets—like potassium, calcium, silicon and iron—were forged in larger stars. All elements more massive than iron, the types of elements upon which technology is reliant—like copper, silver, gold, germanium, tungsten, and many others—were both created, and blasted into the galaxy in supernova explosions.

Some of the material from previous generations of stars recoalesces into new generations of stars, to participate in more rounds of the nucleosynthesis process, some to, again, be blown into the galaxy in stellar explosions. Astronomers call this the star-gas-star cycle.

In elliptical and irregular galaxies, where the gravitational attraction from other stars is more or less isotropic, with no preferred direction, material from supernovae simply disperses homogeneously. This is likely true in the central bulge of spiral galaxies as well. In the disks of spiral galaxies, however, a lot of stars are concentrated into a narrow belt. When stars explode, the material is confined to that narrow band of stars. When stars at the periphery of the disk explode, the gravitational attraction from millions of stars pulls the expanding gas cloud back into the disk, concentrating planet- and life-forming elements.

The "nearby" elliptical galaxy M87 (Fig. 11.6) is a monster, with trillions of stars. Despite that, because of its elliptical geometry, it is less likely to be teeming with life than spiral galaxies, like the Milky Way or Andromeda, a fraction of its size. When galaxies, irrespective of shape, collide, they often merge and form elliptical galaxies. In about 4 billion years, before Sol ends its life, the Milk Way and Andromeda will collide and merge. Although there could be a "brief" (a few million years) period of stellar formation, the final result will likely be a large elliptical galaxy—one where the prospects for future origin and evolution of life is dim.

11.3 Location, Location, Location III: The Right Kind of Planet

The majority of the planets detected in the early days of the exoplanet discovery boom were gas giants, like Jupiter and Saturn in the Solar System. This was due to a methodological bias that made more massive planets much easier to detect than smaller planets, particularly ones that orbit close to their parent star. The first method astronomers developed to detect exoplanets relied upon the fact that when a planet orbits a star, it causes the star to wobble slightly—alternatingly approaching and receding in our line of sight. The bigger the planet, the bigger the wobble. A newer "transiting" method looks for the dip in a star's brightness that occurs when a planet's orbit causes it to pass in front of the star, or transit, blocking some of the star's light[8]. Again, the bigger the planet, the bigger the signal. (Of course, if the plane of a planet's orbit doesn't cause it to pass in front of the star as viewed from Earth, the transiting method won't detect any planets either. This is why scientists have to use statistical methods to extrapolate from the numbers of planets they *do* detect to get the total number of planets actually around.)

Although our knowledge of the nature of stars and stellar lifetimes relegates some famous science fiction planets forever to the realm of fiction, some of the new worlds that are being discovered are so bizarre that many astronomers would have confidently predicted that they couldn't exist until they were, in fact, discovered: planets over 90 percent as old as the universe itself, planets with supersonic windstorms, planets which survived being engulfed when their host star expanded into a red giant, gas giants orbiting closer to their parent star than Mercury orbits to Sol, the list goes on. A good starting point for future science fiction stories would be a simple examination of the real worlds astronomers are discovering every day.

Despite our (potential) disappointments with stars like Gliese 581 and Gliese 667 C, hundreds of Earth-like planets *have* been identified in recent years. "Earth-like," in this context, means a planet having no more than about two times the mass of the Earth, and likely to have a solid surface. So far, all astronomers really know about these rocky planets are their masses, how far they orbit from their host star, and—if they've been found via the transit method—their approximate diameters ("approximate" meaning "within a factor of 2"). Knowing how distant they are from their parent star allows us to make broad estimates about likely surface temperatures, but consider that, in the Solar System, Earth and Venus are in similar orbits (Venus' orbit is about

---

[8] If you're interested, you can now do this kind of planet spotting yourself with a modest telescope, a tracking mount, a digital camera, and the right software.

Table 11.1 Some nearby Sol-like star systems with potentially habitable planets, and the science fiction movies and series in which they or their inhabitants play a significant role

Star system	Distance (light-years)	Movies/tv shows
Alpha Centauri (B the most likely of the three stars)	4.4	*Avatar, Lost in Space, Doctor Who*
Epsilon Eridani	10.5	*Babylon 5, Star Trek, Virtuality*
Gliese 581	20.3	*Gliese 581, Gliese 581 g*
Tau Ceti	11.9	*Star Trek, Barbarella, Earth: Final Conflict*
40 Eridani	16.5	*Star Trek*
Zeta Reticuli (no planets yet observed, but the irregular distribution of a debris disk suggests they are likely to be present)	39.5	*Alien, Aliens, Prometheus, X-Files* (implied)

72 percent as wide as the Earth's) but the average surface temperature of the Earth is 14 °C (57 °F), while, due to a runaway greenhouse effect, the surface temperature on Venus is 462 °C (860 °F).

Most of the newly discovered planets were detected by the Kepler Space Telescope, which began operation in 2009[9]. Kepler was designed to look for transiting planets in a small patch of sky and the ongoing analysis of the data it returned will likely result in hundreds, if not thousands, of planets being added to the roster. About half of the planets spotted by Kepler fall into the size range considered Earth-like. Many of these planets also appear to be in their stars Goldilocks zones around a star. Extrapolating from Kepler's results, astronomers estimate that there could be between 100–400 billion Earth-like planets orbiting within Goldilocks zones in the Milky Way Galaxy. Despite the many planets from science fiction lore that have been relegated to fiction, writers will be tickled to know that there are no shortage of real-universe planets for visits, adventures, even stellar Federation membership (Table 11.1).

Even better, some of the Earth-like planets that have been detected, by either Kepler or ground-based telescopes, are close enough that we might be able to visit them—either in person or vicariously with robotic probes. Such a journey would most likely take centuries, but it's not an impossible engineering challenge, especially for a robot.

---

[9] Kepler's main mission ended in 2013 with the failure of 2 of its four reaction wheels (these are called "gyros" on the Hubble Space Telescope) that are used to point the spacecraft precisely. However, researchers have worked out a way for it still to do some observations with just two wheels.

Still, we should accept these discoveries with the expectation that few of these Earth-like planets will actually be a second Earth (or a "Class M" planet in the terminology of *Star Trek*.) This is why there was particular excitement regarding one planetary system discovered around Kepler 62, a star 1200 light-years away in the constellation Lyra. Kepler 62 seems to have five planets, with two that astronomers believe to be both solid and orbiting within the habitable zone: Kepler 62e and Kepler 62f. Both planets are larger than the Earth, with radii 1.6 and 1.4 times that of Earth, and could well have large amounts of liquid water on their surfaces.

Perhaps the most Earth-like planet discovered to date—at least in solid cross section—is Kepler 78b, orbiting a Sol-like star about 400 light-years away in the constellation Cygnus. The planet is about 1.2 times the radius of Earth, about 69 percent more massive, and astronomers believe that it has a density somewhere between that of Venus and Earth—meaning that its ratio of metal to rock is similar to these two planets.

There are also major differences as well. This planet orbits its sun at a distance that is 1/40th the distance from Mercury to the Sun. It is so close that it has an 8½ hour orbit, and when the star was younger, and slightly larger, the planet (if it was at its current distance) would have orbited within the outer layers of its star. Despite similarities to Earth, the surface of this world is almost certainly molten rock.

Nonetheless, despite the proximity to the parent star, hot rocky planets in the vein of Kepler 78b or Mercury have served as the setting for several stories. These are called Class Y, or "Demon" planets in *Star Trek*. The planet Crematoria in the *Chronicles of Riddick* (2004), home to a "triple max" prison, is such a planet, where the dayside temperature is 700 degrees, while it is 300 degrees below zero on the dark side[10] (Fig. 11.7).

Crematoria is simultaneously an example of where screenwriters set the bar far too low from a "weirdness" standpoint—where the Universe has given us even more extreme worlds—while, at the same time, creating a physically implausible planet. The temperature extremes of Crematoria are on par with those of the planet Mercury, and far more tame than Kepler 78b. So take a planet the size of Kepler 78b, place it at the orbit of Mercury, and that's a close approximation to Crematoria.

---

[10] Immediately obvious is that the Fahrenheit scale has survived to the 28th century. This is just good-natured ribbing. Clearly a film has to be written in the vernacular of the intended audience, and American audiences are still not wholly comfortable with the Celsius scale.

Fig. 11.7 The entry to the "triple max" prison on the hostile planet Crematoria from *The Chronicles of Riddick* (2004). (Copyright © Universal Pictures. Image credit: Movi-eStillsDB.com)

On the other hand, Crematoria does have some issues. Planets as close to their star as Crematoria would very likely be tidally locked[11,12], presenting the same side of the planet to its star at all times. While one might expect that would give the temperature extremes depicted in *Chronicles of Riddick*, the planet is depicted as having an atmosphere. With a 700 degree dayside, most common atmosphere-forming gases would escape into space, leaving only CO_2. CO_2 is a greenhouse gas that traps heat, which would drive up the surface temperature to beyond 700 degrees. On Venus, which has a predominantly CO_2 atmosphere, the gas is efficient at transferring heat planet-wide, so that even though the planet spins very slowly, it is the same temperature everywhere. There is no diurnal variation.

The biggest issue with Crematoria is its breathable atmosphere. Since oxygen is highly reactive, for the atmosphere to be breathable there would have to be something living there to continually resupply it. While certain aspects of Crematoria, like its size, its temperature, and its terrain, are certainly within the realm of plausibility, there are a lot of factors that go into creating an Earth-like planet, even one harsh enough to host a triple max prison. Setting an Earth-like planet in Mercury's orbit would not produce a Crematoria.

To understand just how rare a second Earth could be, it is worth looking at some of the peculiarities that make Earth … Earth. These peculiarities are a large part of what made the Earth suitable for the formation and continued existence of life. For one thing, Earth has a ridiculously huge moon. Large

---

[11] Mercury is not tidally locked, but it spins extremely slowly. Its day is 176 Earth days long, while its orbit is 88 Earth days long. Yes, Mercury has two years in one day.

[12] As was the planet Remus of the Romulan Star Empire, which viewers learn in the film *Star Trek: Nemesis* (2003). The character of Shinzon, a clone of Jean-Luc Picard, says that Reman civilization was confined to the cooler permanently dark side of the planet.

moons are common around gas giants, but they are still tiny relative to the mass of the planets they orbit. Of the terrestrial planets, Mercury and Venus have no moons, and Mars's two moons, Deimos and Phobos, are tiny captured objects, likely from the Asteroid Belt. The Moon, on the other hand, is the largest natural satellite relative to its host planet in the Solar System, with a mass a little over 1 % that of Earth.

Our current best understanding about how the Moon formed is that, in the disordered early Solar System of 4.6 billion years ago, a protoplanet about the size of Mars, dubbed Theia, smashed into the equally proto-Earth. Much of the impactor was incorporated into Earth, and debris from the resulting "splatter" remained in orbit around Earth and eventually coalesced into the Moon. There is an even newer model that suggests that two similarly-sized objects collided to form the Earth-Moon system[13]. In either scenario, born from a cosmic accident, the Moon has subsequently had a huge impact on the natural history of Earth—to the point where, if the Moon did not exist, it is very unlikely life on Earth would exist.

Earth's original atmosphere was vented from volcanic outgassing, and delivered from the outer Solar System in the form of impacts by countless comets. The types of gases that composed the first atmosphere were those similar to the ices commonly found in comets: water, carbon dioxide, ammonia, and methane. Carbon dioxide is the important thing here. If you have ever seen dry ice fog at, say, a Halloween party, that is carbon dioxide. The mass of a carbon dioxide molecule is over twice that of water, ammonia, and methane, and significantly higher than the N_2 (nitrogen) and O_2 (oxygen) molecules that, together, make up 99 percent of Earth's atmosphere today. By virtue of being a more massive molecule, CO_2 is easier for planets to hang onto, explaining why it composes over 95 percent of the atmospheres of Venus and Mars.

The moon-forming impact very likely boiled off Earth's original budget of carbon dioxide and, were it not for that event, our planet would be saddled with an overabundance of the greenhouse gas CO_2. Had that been the case, our planet would have an atmosphere more like the dense blanket that enshrouds Venus, making Earth's temperature several hundreds of degrees hotter, liquid water nonexistent, and life as we know it impossible.

Time passed and volcanic outgassing and further comet impacts combined to form an atmosphere that was CO_2-rich, but significantly less dense than that of Venus. Although largely devoid of oxygen[14], the atmosphere was rife for the formation of life. Early forms of bacteria spread over the face of the

---

[13] Canup R (2013) Planetary science: Lunar conspiracies. *Nature* 504: 27–29. http://www.nature.com/news/planetary-science-lunar-conspiracies-1.14270.
[14] In fact, researchers have shown that because it was devoid of oxygen, the first life thrived—many simple organic molecules do not form in the presence of oxygen. This has dire implications should life

planet and some evolved to take in the atmosphere that Earth provided and exhale a caustic, highly chemically reactive, metabolic toxin (at least for most other bacteria on Earth at the time): oxygen.

Over millions of years, the bacteria, and, later, algae, kept exhaling oxygen—most of which reacted with iron in the crust, methane in the atmosphere, and decaying organic matter. When those oxygen traps became saturated, the levels of oxygen rose[15], dramatically and quickly, killing off many forms of bacteria in a mass extinction event, triggering a major planet-wide glaciation, and leaving Earth with the atmosphere it has today. When somebody professes a belief that human beings are incapable of altering Earth's atmosphere, simply point to what microscopic bacteria did just over 2 billion years ago. Oxygen is such a reactive chemical, that if astronomers ever detect its presence in the atmosphere of other planets in more than trace amounts, it will be a screaming announcement of the presence of life[16].

The moon-forming event had other benefits for life on Earth. The orientation of Earth's axis is not perpendicular to its orbit plane, but rather is has a tilt of around 23°, called its *obliquity*. The Moon prevents Earth's obliquity from changing very much, which stabilizes the nature and variability of the seasons. For comparison, Mars, where the tilt has ranged from 10 to 50° over the course of its history, has experienced dramatic variations of the planet's climate. While Earth has ranged from being far hotter than today to being a global ice cube, those variations are still far less than they could have been had the Moon been absent.

The upshot of all this is that it's easy to imaging many, many worlds that might have started out very much like early Earth, but without a large moon to stabilize the climate, without an impact event to rid the planet of excess CO_2, or without the right kind of bacteria, they would end up becoming places where you couldn't beam down wearing just a (red) shirt. In fact, since CO_2 is so easy for planets to retain, in our quest to find extrasolar Earths, expect to find many more extrasolar Venuses along the way (Gliese 581 C appears to one of the first examples of this).

The reimagined *Battlestar Galactica* had a good handle on the less-than-ideal nature of most "habitable" worlds (unlike the original 1978 show, which had "planet of the week" tendencies). Across the show's 75 regular episodes, once they leave their home worlds, the colonists spend time on only five hab-

---

ever be extinguished from Earth—in the presence of left-over oxygen from the present ecosystem, it is unlikely that life would re-evolve.

[15] This is known, literally, as the "rise of oxygen" or the "Great Oxygenation Event."

[16] Returning to Riddick's universe, with this in mind, one has to wonder how a planet like Crematoria would have evolved a breathable atmosphere in the first place since the dayside temperatures would have caused any liquid water to become gas and, likely, escape the planet.

itable worlds, including New Caprica, with its endlessly lousy weather and poor soil, and the even more marginal "Algae planet," which only just supports some scrubby plant life before getting obliterated by its unstable sun (this was, essentially, Earth towards the end of the rise of oxygen, but orbiting a larger and shorter-lived star). This was a conscious choice by the producers of the show. In the series bible, Executive Producer Ron D. Moore wrote:

> *Galactica's* universe is also mostly devoid of other intelligent life. Unlike *Trek's* crowded galaxy filled with multitudes of empires, ours is a disquieting, empty place. Most planets are uninhabitable. Breathable air and drinkable water are rarities. When we do encounter a world remotely capable of supporting human life, it will be a BIG DEAL. Likewise, an encounter with true alien life will be a HUGE DEAL, and our aliens will not be the usual assortment of bumpy-headed people that are essentially human in all but appearance. Any alien on Galactica must be alien in the truest sense of the word—a creature so foreign to our ways of thinking and living that we may not even recognize it as life at all.

The backstories of *Earth2* and *Firefly/Serenity* (2002/2005) are also predicted on the idea that star systems with any kind of reasonable real estate are rare.

Despite their lack of solid surfaces, don't write off gas giants completely as places that could have real estate suitable for life, or at least places worth a visit. When we look at a gas giant such as Jupiter or Saturn, we are seeing the top cloud layers of a vast atmosphere. Directly exposed to space, the top layers are cold and at very low pressure, but temperatures and pressures rise to hellish levels deeper towards the core of the planet. While the atmosphere is never breathable, there may be intermediate altitudes where the temperature and pressure could allow the construction of floating habitats without requiring super tough materials.

We are on the ragged edge of speculation here, but these habitats might become economically viable by extracting valuable gases from the gas giant, a la Cloud City on Bespin in 1980's *Star Wars: The Empire Strikes Back*. Cloud City extracted the fictional tibanna gas[17], but a real gas that may be of value to a future space economy is helium-3, which makes a great fuel for fusion reactors.

Gas giants tend to have scores of moons, ranging in size from glorified boulders to small planets (Mercury has a diameter of 4879 kilometers, while

---

[17] In the *Star Wars* expanded universe, it is established that when coherent energy, like laser light, passes through Tibanna gas, the gas emits much more energy than that of the original beam. It is an attempt to explain how small hand blasters can emit so much energy. Blasters, in particular hand blasters, are a trope in science fiction, at least they *tried* to offer up a rationale for the extreme amounts of energy per blast.

Ganymede, the largest moon of Jupiter, has a diameter of 5268 kilometers). So its not surprising that large, habitable, moons pop up often in science fiction, such Yavin IV in *Star Wars: A New Hope*, Endor in *Star Wars: Return of the Jedi* (1983*)*, Pandora in *Avatar*, or LV-233 in *Prometheus*.

The existence of an Earthlike moon floating around a gas giant is not without its conceptual rough patches, though. For starters, the moons of the gas giants in the Solar System have a very different composition to the terrestrial planets such as Earth or Mars. Indeed, what a planetary scientist considers a "rock" is dependent upon the context of where it is in the Solar System[18]: at the temperatures found in the outer Solar System, ice is considered to be a rock as it is as hard as granite (when researchers test drills on Earth intended to simulate penetrating the ice of Europa, they use slabs of rock, not lumps of frozen water). Because they formed from a colder outer region of the primordial cloud of gas and dust that gave birth to the Solar System, the surfaces of moons in the outer system are typically composed of ice (Jupiter's moon Io, with its frenetic, all-encompassing, volcanism is an exception, with any ice boiled off long ago—but that makes it even *less* habitable, with a surface that's mostly molten sulfur). What this means is that if you had a moon with an atmosphere at anything close to Earthlike in terms of temperature, you would not have any solid ground, let alone mountain valleys for alien bioweapon research facilities as in *Prometheus*. You'd have a very deep global ocean surrounding a solid small core.

Another difficulty with the existence of an Earthlike moon is that the space around a gas giant is a very dangerous place. Gas giants generate powerful magnetic fields which trap and accelerate subatomic particles to very high energies—the space probes NASA has sent to the jovian planets in our Solar System—such as *Voyager, Galileo*, and *Cassini*—have had to carry along a lot of shielding to prevent their electronics from being fried from this particle radiation. The film *Outland* (1981)—essentially *High Noon* in space starring Sean Connery—was set in a mining colony on Io before science had a firm grasp on just what a harsh radiation environment exists that close to Jupiter, as well as an understanding how long a person, even suited, could withstand that environment (very briefly).

Many gas giants *have* been discovered orbiting much closer to their stars than in the Solar System[19]. Moons too close to the gas planets get irradiated, and computer simulations of moons orbiting these so-called *Hot Jupiters* show that any moon distant enough from its planet to be safe from radiation,

---

[18] To be clear, a "rock" is an aggregate of minerals, and a mineral is any naturally-occurring crystalline substance.

[19] Due to a selection bias in the methods astronomers use to detect exoplanets preferentially detecting large planets close to their parent stars.

would have an orbit that is unstable over long time periods due to gravitational interactions with the parent star.

Because of radiation levels and the icy cold of the outer Solar System, the moons of gas giants were considered de facto uninhabitable for any form of life for a long time. In more recent years the discovery of liquid water under protective layers of ice on several moons of Jupiter and Saturn has led to a reappraisal of their habitability. Moons like Titan, Ganymede, and particularly Europa, appear to have an icy crust atop an ocean of water. On Earth, wherever there is liquid water, there is life, if only microbial, so these sub-surface oceans could be one of the best places to look for extraterrestrial life close to home. This was used as the basis for *Europa Report* (2013), where a private firm funds a crewed mission to Europa, only to find that they may have gotten more than they bargained for.

One moon has even managed to retain a very significant atmosphere: Titan, the largest moon of Saturn and the second largest moon in the Solar System[20]. That atmosphere, like Earth's, is mostly nitrogen, and the atmospheric pressure on the surface of Titan is *greater* than at the Earth's surface—about 1.6 times as high. It's cold and dark though. Surface temperatures average about $-179\,°C$. What little light Titan receives from the distant Sun is further diffused by dense orange smog-like clouds and haze, so the surface gets less than 1 percent of the sunlight that falls on Earth. Nevertheless, there is weather on Titan, with rain falling into rivers and lakes. It is far too cold on Titan for liquid water, so liquid methane takes its place in the evaporation/condensation cycle. Like Europa, Titan may have an ocean of liquid water beneath an icy crust, and may even have volcanoes that spew liquid water.

Even with the cold, dark, and methane, and an icy composition, the surface of Titan vies with Mars as the most Earthlike place in the Solar System. In fact, the environment on Titan is probably very similar to what the Earth itself was like billions of years ago, before microorganisms began terraforming the planet by injecting oxygen into the atmosphere (Fig. 11.8).

Titan had a role in the 1997 film *Gattaca*. Vincent Freeman (Ethan Hawke) is hired by the Gattaca Corporation to plan a mission to Titan. In order to work at Gattaca, who hires only those with the best genetic make-up, Freeman must "borrow" the identity of genetically superior Jerome Morrow (Jude Law). Freeman's description of the smoggy environment of Titan to Morrow, using a cigar and snifter of brandy as props, is a truly inspired scene. Another

---

[20] Titan, like Ganymede at Jupiter, is larger than the planet Mercury. Were they either in orbit about the Sun rather than a planet, there would be no debate over their status—they would unequivocally be classified as planets.

Fig. 11.8 The surface of Titan, as revealed by Huygens, the only probe to land on a moon of the outer Solar System. Huygens was carried to Titan by the Cassini mission in 2005. (Credit: ESA/NASA/JPL-Caltech/University of Arizona)

nice touch is that Uma Thurman's character, the co-worker with whom Freeman strikes up a romantic relationship, is named Irene Cassini.[21]

Hollywood Box: Shortest. Science. Consult. Ever

When Jamie Paglia, showrunner of *Eureka* (2006–2012), was looking for an interesting interplanetary destination for the scientists of *Eureka*'s Global Dynamics to explore in the back half of season four, believing that Mars had been done to death, he called his science advisor (author KRG). The ensuing conversation was one of the shortest science consults in history. Grazier was about to step into the shower, noticed that it was Paglia calling, and decided to answer:

---

[21] Jean-Dominique Cassini, who was born in Italy and was originally Giovanni Domenico Cassini before moving to France, is the astronomer for whom the Cassini spacecraft is named. Cassini discovered four moons of Saturn, as well as the gap in Saturn's main rings, which is known as the Cassini Division (contrary to popular belief, the Cassini Division is not located at the Jet Propulsion Laboratory).

Grazier: Hey Jaime, what's up?
Paglia: What's cooler than Mars?
Grazier: Titan.
Paglia: 'K, thanks! <click>

Thus was born the *Astraeus* Mission to Titan.

Planets and moons aren't the only real estate of value to would-be space colonists, or screenwriters looking for exotic settings for exciting space adventures. There are metallic, rocky, and icy planetesimals—and objects whose compositions are all manners of permutations of those substances, left over from the birth of the Solar System.

Most asteroids are found relatively close to the Sun; a large concentration of which can be found in the asteroid belt that lies between Mars and Jupiter. The asteroid belt is the remnant of a failed planet or two that was likely smashed to pieces, and whose reformation was disrupted by the gravitational influence of Jupiter.

Science Box: Planets at a Glance

Give a planetary scientist two pieces of information—its density and how many craters are on its surface—regarding a planet, a moon, an asteroid, or a comet, and that scientist can instantly tell you a surprising amount about that object. For a screenwriter in the process of world building, these tricks could prove quite handy.

Are you dense?
The density (ρ) of a substance is a measure of how tightly-packed its matter is, and is given in mass per unit volume, either grams per cubic centimeter, or kilograms per cubic meter:

$$\rho = \frac{M}{V},$$

where M is the mass of the object, and V is its volume.

Knowing the *bulk density*, or average density of a planet goes a long way towards determining its composition. Solid planets, and comets and asteroids too, are composed largely of three types of materials: ice, rock, and metal. Ices have a rough density of 1.0 g/cm³, rock is about 3 g/cm³, and metals like iron and nickel weigh in between 7 and 8 g/cm³. The density of Earth is 5.5 g/cm³, and if you had just that information, you could (correctly) assume that Earth had an iron core with a rocky envelope. If a passerby on the street told you that our Moon had a density of 3.3 g/cm³, you could confidently say, "How interesting, no iron core like Earth. Well... maybe a *tiny* one." You learn that the density of Jupiter's moon Io is 3.5 g/cm³; you say, "Rock and probably a small metallic core." If you knew that the largest Solar System asteroid Ceres had a density of 2.1 g/cm³, you would come to the, until recently, extremely surprising conclusion that it was a combination of rock and ice[22]. Clearly, knowing the density of a Solar System object is a very useful fact.

---

[22] Until recently, scientists didn't believe that icy objects without atmospheres could exist as close to the Sun as Ceres (2.8 AU).

"Show don't tell"

It is one of the mantras of a screenwriter: Show don't tell. If it is important enough to get into dialogue, it is important enough to get onscreen. This is closely related to "A picture is worth a thousand words," and in the case of planets and moons, this is doubly so.

An image of the surface of a planetary body is just as powerful as knowing its density in helping scientists quickly determine its nature. The count of the number of craters on a planet or moon allows scientists to determine instantly if that object has an old or young surface. It tells nothing about the age of the object itself, as all the objects in the Solar System are approximately 4.6 billion years old, and formed at nearly the same time.

Over time, objects in the Solar System collide with one another. In the early days, collisions were more common, because impactors were more common. Still, today there are impacts between objects, impacts that leave craters. Still, some objects, take Earth for example, remove their craters. Earth has plate tectonics, Aeolian (air) erosion, fluvial (water) erosion, glaciation, and volcanism. All conspire to resurface Earth to rid it of its craters, signifying that Earth has a young surface. Jupiter's moons Io and Europa are geologically active, are subject to resurfacing, and have few craters (Fig. 11A.1).

If an object like Luna, Earth's Moon, or Mercury, or Jupiter's moon Callisto (Fig. 11A.2) has a wealth of craters, it means that this surface has been exposed to cosmic pummeling for millions, even billions of years, and it has no way of removing these craters. The object has an ancient surface, and no resurfacing mechanism.

Succinctly, more craters means an older surface and a geologically dead object. If an object has few craters, it is a dynamic body, and although a scientist may not be

Fig. 11A.1 Recently resurfaced moons Europa (*left*) and Io (*right*). (Image credit: NASA/JPL/Caltech)

able to tell at first glance what processes are resurfacing the object, a dearth of craters says that *something* is going on here. Show don't tell. For a screenwriter, show an object with few craters, and instantly the information is conveyed to viewers that this is an interesting world that should be either fun or dangerous to explore.

Fig. 11A.2 Moons with ancient surfaces, Luna (left) and Callisto (right). (Image credit: NASA/JPL/Caltech)

Depictions of asteroid belts in science fiction productions typically follow the model of *Star Wars: The Empire Strikes Back*—tumbling mountain-sized boulders fill the screen like a slow-moving flock of birds (the image has enough dramatic appeal that it features even in some non-fiction depictions of the Solar System's asteroid belt, such as in the 2014 remake of *Cosmos*.) In fact, space is still big enough that the asteroid belt is actually very sparsely populated. If you were standing on the surface of most asteroids in the belt, it would be challenging to see the closest other asteroid to you without a telescope—a notable exception being binary pairs like Ida and its moonlet Dactyl.

There is an entire population of near-Earth asteroids in the inner Solar System, ranging from the orbit of Mercury to just outside of the orbit of Mars. Scientists are now starting to understand that many of these objects are actually degassed comets and are, in essence, "rubble piles": loosely-bound rock left over after the icy component of a comet has been boiled off to space.

A steady rain of tiny asteroids and asteroid fragments enters the Earth's atmosphere every day. Most of these asteroids are very small—sand- and pebble-sized objects—and so they are completely burnt up when they enter the Earth's atmosphere, creating short-lived shooting stars. Occasionally a much larger object comes along, such as the 20-m-wide asteroid that appeared in the skies above Chelyabinsk in Russia on February 15, 2013. Because of the shallow angle of entry, that asteroid disintegrated in an airburst almost 30 km above Earth's surface, releasing more energy than 30 of the nuclear bombs that destroyed Hiroshima.

Keeping tabs on the large objects that could, one day, impact Earth is an effort that is getting an increasing amount of attention among astronomers and governments today. The impact scenario has certainly caught the attention of Hollywood, as there have been several films predicated on an asteroid or comet impact scenario: *Deep Impact* (1998), *Armageddon* (1998) (See the science box in Chapter Five "*Armageddon* and The Cretaceous-Paleogene Event"), *Asteroid* (1997), *Meteor* (1979), and *The Day the Sky Exploded* (1958).

To adopt the view that asteroids are merely big stones laying in wait to pummel our planet is to miss a universe of potential. Some asteroids are actually composed of solid metal: primarily a mixture of nickel-iron, but with other metals mixed in. A metallic asteroid with a diameter of one kilometer could yield, among other things, over 2000 tons of gold and 130,000 tons of platinum, as estimated by planetary scientist Jeffrey S. Kargel and others. Other asteroids could be mined for compounds like the water and oxygen chemically bound to various minerals. In the long run, those resources could be more valuable than gold and platinum, since they could be used to supply the raw materials for human exploration and colonization missions in the depths of the outer Solar System.

Consequently, serious proposals have been made to mine asteroids, most notably by Planetary Resources, a startup founded in 2010 and funded by some of the biggest names in technology. Somewhat surprisingly then, especially as it's a pretty common trope in literary science fiction, asteroid mining is relatively rarely depicted on screen. Examples can be found, however, such as the asteroid mining facility depicted in the 2005 *Battlestar Galactica* episode "The Hand of God," in which the Cylons have established an asteroid base to extract and refine the rare mineral tyllium that is used to fuel Cylon and human spaceships.

Asteroid bases, which offer free shielding from cosmic and solar radiation, would also be good places to construct facilities that are too secretive or dangerous to base upon a planet, such as the prison (originally a mining facility) in the 1997 *Deep Space Nine* episode "In Purgatory's Shadow," or the single-occupant prison in the *Space: 1999* episode "Death's Other Dominion." There was also the military outpost located on the asteroid Demon's Run in the 2011 *Doctor Who* episode "A Good Man Goes to War."

Beyond the orbit of Neptune, there is another belt of left-over planet-forming material, the *Kuiper Belt*. The Kuiper belt is much larger than the asteroid belt, with more objects spread over vastly more space. (Pluto is part of the Kuiper belt.) The composition of planetesimals beyond Neptune (beyond Jupiter, really) includes a lot more ice than the asteroids in the inner Solar System[23]. That these kind of objects, as well as icy moons, exist is what makes astronomers snort whenever they watch movies or TV shows in which aliens invade Earth to steal its water, such as the 1983 miniseries *V*. After all, in order to journey all the way to Earth, the aliens must have passed *right by* giant floating blocks of ice, not just once but *twice*, counting the icy planetesimals in their home star system and the ones in ours.

Science Box: What, Exactly, Defines a Planet?

When Pluto was demoted to "dwarf planet" in 2006, after being listed as a planet for 76 years after its discovery in 1930, it was because the International Astronomical Union decided to adopt an all-encompassing definition of what, exactly, a planet was. More to the point, they set out to define the lower boundary between what defines a planet, and what defines "something else."

The size progression from the smallest ice grains in Saturn's rings to stars that, if they replaced Sol, would extend beyond Jupiter, is, in many respects, a continuous spectrum. Human beings, particularly scientists, like to catalog things in Nature, but Nature often bristles against the notions of sharp boundaries and clear distinctions.

At the upper end of the planet spectrum, astronomers initially defined a star as an object with ongoing nuclear fusion in the core, and anything smaller, like Jupiter which is commonly called a "failed star," was something else. Many of the first planets detected around other stars were jovian, even superjovian planets, with some objects far more massive than Jupiter. Jupiter, on the other hand, is about as physically large as a planet can be. Add mass to Jupiter (as in *2010: The Year We Make Contact*), and Jupiter would start to compress under its own mass. Add more, and it would eventually expand again, and by the time you've added enough mass to create an object much larger than Jupiter, you have enough mass for the smallest red dwarf stars.

Want to see Nature laugh? Tell her your classification scheme. That objects may exist that were intermediate between Jupiter and the smallest stars was scientific conjecture as early as 1960, but because telescopes at the time lacked the sensitivity, nothing in this intermediate range was ever detected. In 1975 these theoretical objects were given the name "brown dwarfs" by astronomer Jill Tarter (the real scientist on who *Contact's* Ellie Arroway was based). The first brown dwarfs (Teide 1 and Gliese 229B) were not officially detected until 1995. Today, astronomers know that some brown dwarfs drift independently, while others orbit parent stars.

Some generate energy by fusing deuterium or hydrogen-2 (which is easier to fuse than hydrogen-1), and some even fuse lithium, blurring the lines between previous definitions of "planet" and "star." Although there is debate about these limits,

---

[23] Though scientists are discovering that many asteroids in the outer asteroid belt, beyond 2.5 AU, are icy, while those in the inner belt are "dry."

brown dwarfs are defined to have masses between 13 and 75–80 Jupiter masses. Like Jupiter, they probably have a banded appearance.

Astronomers and planetary scientists never defined a lower limit, because there was no real need, and pretty much stuck to the "If it walks like a duck and quacks like a duck" definition. That changed in January 2005. If the upper limit definition of "planet" seems like a can of worms, that was simplistic to the events that occurred when scientists attempted to define the lower limit of "planet."

In the 1950s, U.S. astronomer Gerard Kuiper, and Irish astronomer Kenneth Edgeworth independently postulated that, beyond Pluto, is a belt of icy objects that, today, is known as the *Kuiper Belt*, or, more correctly, the *Edgeworth-Kuiper Belt*. In 1992 the first object in this class was discovered, object 1992 QB1[24]. Many more objects were detected, and this was fine, until January 2005 when astronomers discovered an object 27 percent more massive than Pluto. Did we really want our roster of Solar System planets to fill endlessly with frozen lumps of rock orbiting in the Kuiper belt?

In August 2006, the International Astronomical Union (IAU) met to formalize the definition of a "planet." The definition they adopted[25] was:

1. A "planet" is a celestial body that (a) is in orbit around the Sun, (b) has sufficient mass for its self-gravity to overcome rigid body forces so that it assumes a hydrostatic equilibrium (nearly round) shape, and (c) has cleared the neighbourhood around its orbit.

Objects that had failed to clear their orbit were considered dwarf planets.

2. A "dwarf planet" is a celestial body that (a) is in orbit around the Sun, (b) has sufficient mass for its self-gravity to overcome rigid body forces so that it assumes a hydrostatic equilibrium (nearly round) shape, (c) has not cleared the neighbourhood around its orbit, and (d) is not a satellite.

The IAU's decision soon became probably the most popularly debated scientific definition ever, with school children writing aggrieved letters to astronomers in defense of Pluto's former status, but the proponents of the new definition prevailed by pointing to the spate of large objects discovered orbiting beyond Neptune.

However, since then the number of discovered exoplanets has exploded. Consequently, writing in the August 2013 issue of *Sky & Telescope*[26], astrobiologist David Grinspoon argues it is time to revisit the IAU's definition. After all, under the IAU's 2006 definition of a planet, *none* of the objects discovered orbiting distant stars were officially planets. The IAU's definition of a planet is limited to the Solar System because it (a) refers explicitly to Sol, not any other star, and (b) it is currently impossible to detect whether or not a planet in another star system has cleared its orbit. For that matter, there are asteroids that cross the orbits of all the terrestrial planets. Do these planets get demoted? Jupiter has asteroids called the Trojan Asteroids co-orbiting roughly 60 degrees ahead and behind in its orbit—a number of objects on

---

[24] Objects like this have an odd nomenclature. The number refers to the year the object was discovered. The first letter refers to the two-week period of the year in which the object was discovered (52 weeks, 26 letters), the second letter the day (so it's capped at "G"), and the second number, is an order of objects discovered that day.

[25] IAU 2006 General Assembly: Result of the IAU Resolution votes: www.iau.org/static/archives/releases/doc/iau0603.doc.

[26] http://funkyscience.net/wp-content/uploads/2013/07/PlanetDef_Aug.pdf.

par with those of the Asteroid Belt. Has it cleared its orbit? There is a class of objects called Centaur objects between Jupiter and Neptune that cross the orbits of these objects. The orbit of Pluto crosses that of Neptune. In short, none of the planets in the Solar System have cleared their orbits, hence none strictly qualify as planets per the IAU definition.

To complicate matters, there is a new class of planets recently discovered: rogue planets. Starless planets, wandering adrift, have appeared in science fiction, in *When Worlds Collide* (1951), and there were probes constructed to explore the rogue planets Ultra and Meta in *Space: 1999* (1975–1976). So far, only a handful of these objects have been detected, the closest being WISE 0855–0714 at 7.1 light-years away, but estimates suggest that there could be as many as 400 billion of these lonely planets adrift in the galaxy[27]... except that, by the IAU definition, they're not planets, either, because they do not orbit Sol.

The debate that cost Pluto membership in the Planet Club was not about Pluto, but about what defines a planet, particularly at the lower limit. What happened to Pluto closely mirrors what happened to Ceres a little over 200 years earlier. When Ceres, the largest asteroid[28] in the main belt between Mars and Jupiter, was first discovered, it was originally classified as a planet, because there was no precedent to name it anything else. As more and more similar, smaller, objects were discovered, they were given their own class. When scores of Pluto-like objects were detected, they were given their own class. Clearly the definition of "planet" needs to be re-visited, because, by the IAU definition, the number of planets in the entire Universe is: precisely zero.

It must be emphasized at this point that in recent years both theoretical and observational astronomers have also come to realize that there are many more ways to arrange planetary systems than the model embodied by the Solar System, where a single star is surrounded by an inner system composed of rocky and metallic terrestrial planets, with jovian ice and gas giants relegated to the outer system.[27, 28]

In the original 1978 *Battlestar Galactica*, the Twelve Colonies were planets that orbited one of two suns in a binary star system. At the time, even such an arrangement would have been considered an excess of imagination by some astronomers[29]. They believed that gravitational instabilities generated by the interaction of two stars would prevent planets from being formed—or if they *did* form, the instabilities would either eject any planets into interstellar space or constantly perturb their orbits, making conditions so erratic as to be uninhabitable.

Computer simulations have shown that two types of planetary orbits are, in fact, stable over long time periods in binary star systems: P-type orbits and S-

[27] Sumi T et al. (2011) Unbound or distant planetary mass population detected by gravitational microlensing. Nature 473(7347): 349–352.

[28] Asteroid coined 30 years earlier than known: 'Big data' impacts history of science: http://thedailyjournalist. com/uncategorized/asteroid-coined-30-years-earlier-than-known-big-data-impactshistory-of-science/.

[29] http://history.nasa.gov/CP-2156/ch2.5.htm.

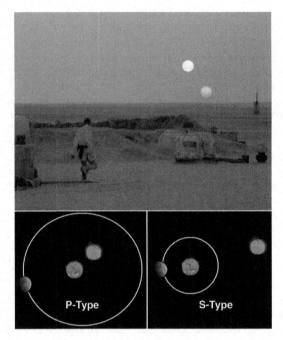

Fig. 11.9 Luke Skywalker (*Mark Hamill*) and he twin setting suns of Tatooine (*above*). Illustrations below show a P-Type orbit (*where a planet orbits a binary pair at a distance*), and an S-Type orbit (*where a planet orbits closely to one member of a binary pair*) that could reproduce Luke's view. (Copyright © Twentieth Century Fox. Image credits: NASA, MovieStillsDB.com)

type orbits (Fig. 11.9). In P-type, or *circumbinary*, orbits, a planet orbits both stars at a distance—to the point where their combined gravitational attraction appears almost like a single source. In S-type orbits, a planet orbits one star so closely that it barely feels the gravitational perturbation of the other star.

One such computer simulation[30] showed that the orbits of planets in S-type orbits around either of the larger yellow stars in the nearby Alpha Centauri system would be stable out to about 3 AU—in the Solar System that would extend from Sol to just shy of the outer edge of the Asteroid Belt, and is well more distant than the outer edge of the Goldilocks zone for either star. As is often the case in astronomy, objects predicted in theoretical or computational models are later found to exist in nature, and binary systems with planets orbiting one or both stars have been detected—in both S-type and P-type orbits. In particular, there is evidence[31] that a Mercury-like planet (only far hotter) orbits very close to Alpha Centauri B.

---

[30] Wiegert PA, Holman MJ (1997) The stability of planets in the Alpha Centauri system. *The Astronomical Journal* 113(4): 1445–1450.

[31] Dumusque X, Pepe F, Lovis C, Ségransan D, Sahlmann J, Benz W, Bouchy F, Mayor M, Queloz D, Santos N, Udry S (2012) An Earth mass planet orbiting Alpha Centauri B. *Nature* 490(91): 207–211.

One implication of all of this is that the iconic image from *Star Wars: A New Hope*, that of Luke Skywalker contemplating his future against the backdrop of the twin setting suns of Tatooine is not just a flight of fancy, but a scientifically plausible scenario—both system configurations in Fig. 11.9 would allow for an observer on the planet to enjoy twin setting suns.

The home planets of the migrating aliens in the TV series *Defiance* (2013–) is also a binary system, with three habitable planets divided between the Sol-like stars Sulos and Vysu. This provided ample room for the establishment of seven sentient species in the system—together known as the Votanis Collective—prior to their migration to Earth.

Alpha Centauri is actually a trinary system—with Sol-like stars Alpha Centauri A and B in mutual orbit, with little red Proxima Centauri, or Alpha Centauri C, in a P-type orbit around both of the larger stars. A similar system, one we have discussed earlier, is Gliese 667. In that instance, a red star is in a P-type orbit around two orange stars. The exoplanets in that system orbit the red star. This is the typical configuration for a trinary system—a small star orbits two larger stars in a P-type orbit[32].

Multiple star systems with even more component stars exist, and tend to be groups of binary pairs that are, in turn, all gravitationally bound to one another. For example, a favorite target at summer star viewing parties is Epsilon Lyrae, the famous "double-double." This system is actually composed of four stars grouped as gravitationally-bound binary pairs[33]. The star Castor, the head of one of the twins of Gemini, is actually a system of six stars—three gravitationally-bound binary pairs.

Given the nature of multi-star systems, as groupings of binary pairs, science only has to show that planets are stable and habitable in a binary system for them to be stable and habitable in systems with many more stars. This means that the Twelve Colonies of the reimagined *Battlestar Galactica* (2004–2009) and its prequel *Caprica* (2009–2010)—with twelve habitable planets in a four-star binary-binary system, much like Epsilon Lyrae—is not much of a scientific stretch. Neither is Kalgesh from *Nightfall* (a famous science fiction novel by Isaac Asimov and Robert Silverberg, adapted for film in both 1988 and 2000). In *Nightfall*, Kalgesh orbits one star in a six-star system, much like Castor.

---

[32] The trinary star system initially introduced in Kara Thrace's vision in the *Battlestar Galactica* episode "He that Believeth in Me" was a bit of an astronomical red herring suggested by the show's science advisor (KRG). With the show entering its final season, fans would be looking for hints that the rag tag fleet was nearing Earth. Astronomically astute viewers might figure that the star was the Alpha Centauri system, the nearest star system to the Solar System, which is a trinary. Instead, the stars were just another trinary system they passed along the way. Based upon comments on Internet threads, the misdirection worked.

[33] There actually appears to be a fifth component to this system, which means that a more accurate description would be of an Alpha Centauri-like system in mutual orbit with a binary pair.

For astronomers, the discovery of exotic star systems and planets has only just begun. For clever screenwriters and novelists, there is no end to the stories—of first contact scenarios, interplanetary wars, planeto-political intrigue, and stellar federations—that could arise from the knowledge that multiple star systems are capable of hosting many habitable planets.

11.4 If We Can't Find a Nice Place, Let's Build One

If our characters are explorers hoping to find a place for an extended stay—and after travelling tens or hundreds of light-years across space any explorers would probably want to—then it would be nice to make things as pleasant as possible: at least enough to grow some food and walk around without an environmental suit.

One way to do this might be *terraforming*, which involves geological and atmospheric engineering on a planetary scale. In the movie *Alien*, the crew of the space tug *Nostromo* are sent to investigate a distress signal on the planet LV-426. A murky, toxic atmosphere with high winds forces the crew to wear space suits whenever they venture onto the surface. By the events of *Aliens*, set about 60 years later, a small group of colonists have been running fusion-powered atmospheric processing stations long enough to make the atmosphere breathable, if only just[34]. In *Firefly* and *Serenity*, terraforming was the ultimate explanation for injustices that motivated a system-wide civil war, one where richer, more successfully terraformed planets dominated poorer, less-developed worlds.

The enormous machinery of *Aliens'* processing stations is cinematically spectacular, but taking a page from Earth's own natural history, real terraforming projects would likely employ bacteria as microscopic solar-powered atmospheric processing plants. This is the premise of *Red Planet* (2000), where a team is sent to Mars to find out why the algae sent to seed the planet and oxygenate the atmosphere have died off[35].

Using biological systems to aid in terraforming is also a critical element of the premise of *Defiance*. Ironically, in this show, it's the *Earth* that's been subjected to transformation, following the botched handling of the arrival of a migrant fleet of aliens. War breaks out between humans and the arriving Votanis Collective, during which terraforming ships crash to Earth. These

---

[34] *The Arrival* (1996) was a sort of role-reversed *Aliens*, with aliens setting up atmospheric processing plants on Earth to make our atmosphere hospitable for them. The film starred Charlie Sheen who is, some would claim, proof that alien life exists on Earth.

[35] In *Red Planet* the algae has been eaten by insect-like creatures that one character refers to as "nematodes." In fact, a nematode is a type of roundworm.

haphazardly introduce many new species of flora and fauna and alter the terrain and climate, creating a hybrid Earth-alien ecosystem in which neither the natives nor the newcomers feel at home[36].

For the truly visionary world builder, re-engineering an existing planet may lack a certain *je ne sais quoi*. They make the leap to constructing structures from scratch, dreaming up habitats that are nothing like traditional planets. One such example is the *Dyson Sphere* (named after its inventor, the physicist Freeman Dyson), featured in the 1992 *Star Trek: The Next Generation* episode "Relics." A Dyson Sphere means taking all the planets and asteroids in a solar system, grinding them up and reconstituting them as a thin shell of super strong material. This shell completely encloses the sun at distance right in the middle of the Goldilocks zone. The end result is a vast interior living space—for our Sun's Goldilocks zone, it would have the equivalent surface area of 550 million Earths. In "Relics," the Dyson Sphere is found abandoned because the central star has become unstable, but there are other problems that make such an enclosing shell problematic, not least the fact that there would be no apparent gravity to hold the inhabitants, or their atmosphere, in place on the inner surface of the shell.

A *slightly* less ambitious idea was proposed by Larry Niven in his 1970 book *Ringworld*, which, at the time of writing, is under development as a TV miniseries for US television. Instead of a complete shell, a thick band is constructed at the appropriate distance instead. The band is set spinning to simulate gravity via centrifugal force, and high walls at the top and bottom rims of the band prevent the atmosphere from spilling over the sides.

Still, having a star at the center of a ringworld introduces orbital instabilities, and the station-keeping thrusters necessary to mitigate these instabilities would be gargantuan, and would require colossal amounts of propellant. A more practical idea, then, is to create smaller structures, similar in concept, called *orbitals*. These are basically scaled-down ringworlds—spun for centrifugal force to simulate gravity and having high rim walls to retain the atmosphere—which would orbit a sun or a planet, instead of completely filling an orbit around a sun or planet. Even at their much reduced size they would still offer vast tracts of livable real estate.

Much of the action of the *Halo* video games franchise (which provides the context for the 2012 *Forward Unto Dawn* live action web series) occurs on various orbitals or "halos." A very bijou orbital above Earth served as the home of the super wealthy in 2013's *Elysium*. The trademark of an orbital—the lack of a "roof" to keep air in (which would be required on a smaller space station)—is what allows a rogue spacecraft from Earth to land directly

---

[36] Since terraforming literally means "Earth forming," the Earth of *Defiance* was, more correctly, Vota-formed.

on someone's lawn without all the traditional mucking around with docking at an airlock (though the halos of *Halo* do have outward-facing airlocks for spacecraft not designed to enter atmospheres).

Those artificial worlds are frequently as detailed and well-thought-out as many science fiction planets. Showrunner Kevin Murphy explains why some much time is spent on the design of these environments: "In *Defiance* we're really trying to build in as much detail as we can from Moment One. We really tried to create a universe where, once you take that leap that 'I'm going to watch that show,' you feel that it's an immersive, well thought-out, well-researched universe." A chink in the armor of the world, or lazy world building, can elicit the dreaded "Oh, please!" moment just as readily as any other technical gaffe.

Science Box: Some Physics of Miller's Planet in Interstellar

In the film *Interstellar*, explorers searching for a new home that would allow humanity to evacuate an environmentally-ravaged Earth discover three habitable candidate planets orbiting a black hole 10 billion light years away—a black hole that has been named Gargantua. The planets orbiting Gargantua were unofficially named for the scientists who first explored them: Mann's Planet, Miller's Planet, and Edmund's Planet.

The film explores many implications of the way relativistic effects would impact life on a planet orbiting within the intense gravitational field of a black hole. Let's examine some of the details and implications that the film simply did not have time to explore. Now, in the film, Dr. Brand (the elder) explains that Gargantua is a rapidly-spinning black hole; this complicates matters greatly by adding additional relativistic effects such as a phenomenon called *frame dragging*. Our calculations will be for the simple case of a non- or slowly-spinning black hole.

How big is Miller's planet? The film gives us a clue. We do not know the mass of Miller's Planet, but we do know from the film that the acceleration due to gravity on its surface was 130 percent that of Earth, which would make it 12.7 m/s². We can plug this into Newton's *Law of Universal Gravitation*, which relates the gravity on an object's surface to its mass, M, and radius, r, of the object, using a conversion factor known as the *gravitational constant, G*, which has a value of 6.674×10^{-11} m³/(kg s²). (G is a tiny number, which is why you need a planet's worth of mass just to hold something to the ground as strongly as a small magnet could.):

$$g_{\text{Miller}} = \left(-\frac{GM_{\text{Miller}}}{r^2_{\text{Miller}}} \right) \approx 12.7 \frac{\text{m}}{\text{s}^2}.$$

For our next step, remember that the mass of an object is equal to the product of its density and volume, and (for a spherical planet) volume is related to the radius in a straightforward way. Substituting the expression for density and volume in place of the mass and doing some basic algebra gives us an equation with just the density and radius (The equation above has a minus sign in front of it because gravity is a force that causes things to get closer together, not increase in distance. By convention g is always measured "downward," so we can cancel out and drop the minus sign.)

$$g_{\text{Miller}} = \left(-\frac{\left(G\left(\rho_{\text{Miller}} \dfrac{4}{3} \pi r_{\text{Miller}}^3 \right) \right)}{r_{\text{Miller}}^2} \right) = \left(\frac{4\pi G \rho_{\text{Miller}} r_{\text{Miller}}}{3} \right),$$

Where ρ_{miller} is the density of Miller's planet, and r_{miller} is its radius. If we assume that Miller's Planet is a terrestrial planet—an iron core surrounded by a rocky envelope[37]—then its density will be on par with those in the Solar System. Since Earth is closest in size to Miller's Planet, assume a similar density as Earth, which is 5.52 g/cm³, or 5520 kg/m³. Rearranging the equation to solve for r_{miller}, we have:

$$r_{\text{Miller}} = \left(\frac{3 g_{\text{Miller}}}{4\pi G \rho_{\text{Miller}}} \right) = \frac{(3)\left(12.7\dfrac{m}{s^2} \right)}{(4\pi)\left(6.67\times10^{-11} \dfrac{m^3}{kg\cdot s^2} \right)\left(5520 \dfrac{kg}{m^3} \right)},$$

or,

$$r_{\text{Miller}} = \frac{\left(38.1\dfrac{m}{s^2} \right)}{\left(4.62\times10^{-6} \dfrac{1}{s^2} \right)} = 8.66\times10^6\,m = 8235\,km,$$

The radius of Miller's Planet, assuming a density similar to Venus, is 122 % that of Earth.

Miller's planet is not close enough to Gargantua to be tidally disrupted—torn apart by gravitational forces. This is one counterintuitive oddity of supermassive black holes: their event horizons are so far away from their centers that the gravitational gradient is fairly small outside the event horizon. In fact, for very large black holes, objects could even pass within the event horizon and not be destroyed. For less massive black holes, objects like planets or spacecraft, would be torn apart long before reaching the event horizon in a stretching process aptly called *spaghettification*.

By the same token, Miller's planet would certainly be tidally locked, with the same hemisphere permanently facing Gargantua. Most large moons in the Solar System, even Earth's, are tidally locked, and continually present the same face towards the planet they orbit. This is why, contrary to popular wisdom[38], there is no permanently "dark side of the Moon"—there is a 'near side" and "far side" relative to Earth. In

---

[37] With a rocky envelope surrounding a metal core, the global ocean adds but a trivial amount of mass.
[38] And Pink Floyd.

the case of Gargantua, planets would have a permanent day and night sides, as illumination would come from the brightness of the black hole's accretion disk.

In the film Dr. Brand (the elder) says that Miller's Planet is so deep within Gargantua's gravity well, that as 1 hour elapses on the planet, seven years elapse on Earth due to an effect known as *gravitational time dilation*. The rate of a slow-running clock immersed deep within Gargantua's gravity well (t_s) compared to a fast-running clock at a great distance (t_f) is given by:

$$t_s = t_f \sqrt{1 - \frac{2GM}{r_{orbit}c^2}},$$

where r_{orbit} in this equation is the distance from Gargantua, and M is the black hole's mass. One hour compared to seven years is a time dilation factor of approximately 61,320, so:

$$\frac{1}{61320} = \sqrt{1 - \frac{2GM}{r_{orbit}c^2}} = S,$$

where S is the "slowdown" factor. (Note the similarities these equations have to the time dilation equations for moving objects in Chapter Nine.) In Chapter Seven, we gave the radius of the event horizon as:

$$R_s = \frac{2GM}{c^2},$$

Substituting, we have:

$$S = \sqrt{1 - \frac{R_s}{r_{orbit}}},$$

Rearranging and using the value of S stated in the film gives us:

$$r_{orbit} = R_s \left(\frac{1}{1-S^2} \right) = R_s \left(1.00000000027 \right),$$

In Chap. 7, we calculated the event horizon radius for Gargantua to be 2.96×10^8 km, or just under 2 AU.[39], yielding an r_{orbit} that is within 80 meters of the event horizon in order to achieve that degree of time dilation.

This has several bizarre implications. If the orbital distance of Miller's Planet varied by more than one part in 3.76 billion, the planet passes through the event horizon and is lost forever[40]. Given that the planet would likely be constantly pummeled by infalling debris, this would be an extremely difficult orbit to maintain so precisely.

---

[39] Typically, you can do the kinds of back-of-the-envelope calculations like we've been doing by calculating out to only two or three decimal places. If you're going to try to recreate the kinds of calculations we do in this science box, you'll need twelve decimal places to do it well. The calculation may be difficult using most hand calculators, but spreadsheets like Excel work well.

[40] In a recent paper, physicist Stephen Hawking says that information—mass and energy—that passes through the event horizon of a black hole may not be lost forever, although it will be jumbled up. This is an area of active ongoing debate, so stay tuned.

If the time dilation stated in the film was that for the planet center, then that places the part of the planet closest to the black hole well within the event horizon. If the time dilation stated in the film was for the point of the planet closest to Gargantua, then what would be the time dilation due to the black hole on the point most distant, or one planetary diameter father away? The equation for that would be:

$$S = \sqrt{1 - \frac{R_s}{\left(r_{orbit} + 2r_{Miller}\right)}},$$

which is a factor of 134. So for each hour that passed on Earth, only 5 ½ days pass on the far/dark side of Miller's planet. In fact, if a six-foot-tall (1.83 meters) astronaut stood at the spot closest to Gargantua, clocks at that person's head and feet would vary by about 4 days over the span of 1 year.

The debris in the accretion disk, as well as the high amounts of thermal radiation the disk would emit, would be major impedances to the ability of any life to survive on the surface of Miller's Planet. That said, Gargantua's accretion disk is depicted with a radius far less than that of the orbit of Miller's Planet, but if Miller's Planet orbits a mere 80 meters from the event horizon, then the entire accretion disk would have been within the event horizon. How did enough, or any, radiation escape the gravity of Gargantua to illuminate its planets? For that matter, when electromagnetic radiation passes from a region of higher gravitation (with a slower ticking clock) to a region of lower gravitation (with a faster clock), the frequency is decreased, and the wavelength increased. This would shift visible radiation to the red end of the spectrum, a phenomena known as *gravitational red-shifting*. With the degree of time dilation stated for Miller's Planet, this would shift the frequency of radiation, emitted or reflected, from the planet to frequencies well outside of the visible range of the spectrum. Restated, Miller's Planet would have been invisible until an observer was practically right on top of it. The frequency of the "pings" Miller sent from the surface of the planet would also have been shifted to the point of undetectability, unless the NASA people back on Earth who were listening for the signal were searching a vastly different frequency range than the frequency with which the ping was sent.

Even ignoring the global ocean and colossal waves depicted in the film, Miller's Planet is truly one of the most bizarre to appear in cinematic science fiction to date. Given narrative constraints, the bizarre implications of existence on Miller's planet were only teased in the film. While the planet would have made an awful future home for the human species, it certainly would have been the subject of intense scientific scrutiny for decades.

11.5 Creating the World is the First Step, Now We have to Explore it

Rockne S. O'Bannon, creator of many science fiction television series (including *seaQuest DSV, Farscape*, and *Defiance*), says that his science advisor on *seaQuest*, oceanographer Dr. Robert Ballard, once told him that the exploration of our oceans will be done increasingly "hands off":

Fig. 11.10 The alien moon Pandora from *Avatar* (2009). (Copyright © Twentieth Century Fox. Image credit: MovieStillsDB.com)

Much of the direct, personal, exploration that, normally, in the show we would want to have our people do, the reality was—especially 25 years into the future—that they would probably send some sort of incredibly sensitive probe, maybe even instead of the ship, and not be threatened at all … and still get the exact same experience.

However, because we're a TV show, and we wanted to have our people hands-on exploring we didn't do that. By extension, every time I now watch *Star Trek*, I always remember what Ballard said. Obviously, multiple centuries into the future, clearly we're not going to be sending ourselves out in ships—with a crew that's vulnerable—to do the exploration. We'll be sending probes and, perhaps, some sort of anthropomorphic robot or something. We won't be necessarily putting ourselves in harm's way.

What is logical, and what works, in the real world, though, is not what keeps the attention of an audience in a dramatic production. O'Bannon continues, "However, having said that, obviously on *Star Trek*, it ain't drama unless Captain Kirk and Mister Spock are out there in great personal peril." So screenwriters will continue to put their characters directly in harm's way, and that means creating new worlds, both natural and artificial, with an emphasis on the exotic, breathtaking, and perilous (Fig. 11.10).

Kevin Murphy believes that, "Sometimes science can be so wildly 'out there' that if you weren't reassured that this is real, it would sound like [BS]." Perhaps there is no field where this belief is better supported than with the strange new planets that astronomers continue to discover on a near-daily basis. Although, in recent years, science has put the kibosh on many beloved planets from days of Hollywood's science fiction past (farewell, lush jungles of Venus), some of the real newly-discovered planets are stranger than anything writers have envisioned to date, and some of those will become prime real estate for adventures. Screenwriters will find ways to make those strange new real-life worlds, and many more still to be discovered, characters in new dramas, just like Hoth, Arrakis, and Pandora.

> The important achievement of Apollo was demonstrating that humanity is not forever chained to this planet, and our visions go rather further than that, and our opportunities are unlimited.
>
> Neil Armstrong, Astronaut

12
Afterword

When I was six years old, my parents called me into the living room and sat me down. "Michael, we think you'll like this," my mom said, turning on the television. It was an episode of the original *Star Trek*. I did like it. I liked it a lot.

That show kindled an enduring and consuming interest in both space and science fiction. Today I'm a professor of astronomy of at the University of Wyoming, following my passion for science, as well as a science fiction writer with a couple of novels professionally published that highlight modern astrophysics. The power of Hollywood to inspire should not be underestimated.

Becoming a member of the editorial board of Springer's Science and Fiction series was a way for me to bring together science and science and science fiction, which can be a powerful mixture for good in the world, educating in addition to inspiring and entertaining. When thinking about who would be an ideal contributor to the series, I immediately thought of Kevin Grazier.

I first met Kevin at a World Science Fiction convention where we shared our mutual interests in science and science fiction. I later invited him to be a guest instructor at the annual Launch Pad Astronomy Workshop for Writers, which I run out of the University of Wyoming. At Launch Pad we teach primarily science fiction writers a crash course in astronomy and the communication of scientific concepts to general audiences. Our goal is to give these writers the basics and encourage them to include more and better science in their work to educate and inspire. Launch Pad alumni include best-selling authors, television writers, video game writers, editors and creators of all types, with a combined audience numbering in the tens of millions. We want to change the world, at least a little bit, into a more science-friendly place, and to help share how amazing our universe actually is.

Our reality is far from boring! Time and space really do warp. Stars can and do explode. Black holes exist. There's plenty of the fantastic for a well-informed science fiction writer to choose from without resorting to fantasy. Besides, some of the biggest fans of science fiction are scientists, and getting the science wrong risks their suspension of disbelief and the loss of an audience.

Kevin was a perfect guest instructor for Launch Pad, he had years of experience as a science consultant for a number of high-profile television shows and movies, and had also co-authored *The Science of Battlestar Galactica*. When I queried him about ideas for the Science and Fiction series, he came back with a proposal for *Hollyweird Science*, which he would write in collaboration with Stephen Cass.

Their pitch epitomized my vision for Science and Fiction, and Springer agreed.

Not only was an exploration of the science of Hollywood television and movies ideal, but the subject was also big. Perhaps not as big as the universe, which is "really big," to quote Douglas Adams, but big enough for more than one volume. I'm happy to announce that there's more Hollywood weirdness coming, with more science meeting science fiction, in a sequel to this volume: *Beyond Hollyweird Science*. I still like it. I like it a lot.

Michael Brotherton
Laramie, WY

Appendices

Appendix A. Film References

Franchise or film series	Film title	Year(s)	Chapter(s)
	The Abyss	1989	3
	Adam	2009	3
	Akira	1988	3
	Alice in Wonderland	1933; 2010	4
Alien	Alien	1979	1, 9, 11
Alien	Aliens	1986	4, 9, 11
Alien	Prometheus	2012	9, 11
	Alien Nation	1988	2
	Aliens of the Deep	2005	3
	AM 11:00	2013	3
	The Andromeda Strain	1971; 2008	2
	Angels and Demons	2009	5
	Apollo 13	1995	3, 5
	Armageddon	1998	5, 11
	The Arrival	1996	11
	Asteroid	1997	11
	Avatar	2009	2, 3, 9, 11
Back to the Future		1985–1999	3
Back to the Future	Back to the Future	1985	9
	Barbarella	1968	11
Batman			3
	Beginning of the End	1957	6
	Behemoth, The Sea Monster	1959	6
	Bill and Ted's Excellent Adventure	1989	9
	The Black Hole	1979	7
	Chain Reaction	1996	2, 4

Franchise or film series	Film title	Year(s)	Chapter(s)
	Chronicles of Riddick	2004	12
	Close Encounters of the Third Kind	1977	9
	The Color Purple	1985	6
	Contact	1997	1, 3, 6, 11
	Coraline	2009	10
	The Core	2003	2, 3
	Creation	2009	2, 3
	The Day After Tomorrow	2004	3
	The Day the Earth Stood Still	1951, 2008	3
	The Day the Sky Exploded	1958	11
	Deep Impact	1998	11
	The Dish	2000	3
	Dune	1984; 2000	9, 11
	The Effect of Gamma Rays on Man-in-the Moon Marigolds	1972	6
	Elysium	2012	11
	Ender's Game	1985	9
	Enemy Mine	1985	2
	Eraser	1996	6
	Europa Report	2013	11
	Event Horizon	1997	7, 9
Fantastic Four			6
Fantastic Four	Fantastic Four	2015	3
	Fantastic Voyage	1966	7
	The Fifth Element	1997	4
	The Fly	1958; 1986	7
	Forbidden Planet	1956	7
	Frequency	2000	10
	Gamera	1965, 1995	6
	Gattaca	1997	11
	Gliese 581	2014	11
	Gliese 581g	2013	11
	Godzilla	1954; 1998; 2014	3,6
	Gravity	2013	1, 2, 4
	The Green Mile	1999	6
	Gugusse et l'Automate	1897	2
Harry Potter		2001–2011	10

Franchise or film series	Film title	Year(s)	Chapter(s)
	The Heroes of Telemark	1965	4
	Hollow Man	2000	3
	Homonculus	1916	2
	Honey I Shrunk the Kids	1989	7
	Hot Fuzz	2007	1
Hulk			6
	Hulk	2003	2, 4, 6
	Hugo	2011	2
	I Am Legend	2007	2
	I Am Omega	2007	2
	I, Frankenstein	2014	3
	I, Origins	2014	3
	I, Robot	2004	4
	The Incredible Shrinking Man	1957	7
	Independence Day	1996	2
	Interstellar	2014	6, 7, 8, 9, 11
Iron Man			3
James Bond	Moonraker	1979	2
James Bond	The Spy Who Loved Me	1977	2
James Bond	The World is Not Enough	1999	3
	John Carter	2012	4
	The Johnstown Flood	1926, 1946, 1989	2
	Journey to the Far Side of the Sun	1969	10
Jurassic Park			3
Jurassic Park	Jurassic Park	1993	2
Jurassic Park	The Lost World: Jurassic Park	1997	3
	King Kong	1933	6
	Knock on Any Door	1949	8
	K-Pax	2001	9
	La Charcuterie mécanique	1895	2
	L'Arrivée d'un Train en Gare de la Ciotat	1895	2
	The Last Man on Earth	1964	2
	The League of Extraordinary Gentlemen	2003	3
	Le Voyage Dans La Lune	1902	2
	Logan's Run	1976	3

Franchise or film series	Film title	Year(s)	Chapter(s)
	Looper	2012	9, 10
	Lost in Space	1998	9
	The Lost World	1925	6
	Lucy	2014	2
	The Man with the X-Ray Eyes	1963	6
Matrix			7, 10
Matrix	The Matrix	1999	5
	Meteor	1979	11
	Metropolis	1927	3
	Nightfall	1988; 2000	11
	The Nightmare Before Christmas	1993	6, 10
	The Omega Man	1971	2
	On the Beach	1959	2
	Outland	1981	9, 11
	Pacific Rim	2013	3, 6
	Planet of the Apes	1968	9
	Predator	1987	6
	Rain Man	1988	3
	Raising Genius	2004	3
	Real Genius	1985	3
	Red Planet	2000	3, 4, 11
	Serenity	2005	11
	Sky Captain and the World of Tomorrow	2004	3
	Sneakers	1992	7
	Solaris	2002	3, 5
Spider-Man	The Amazing Spider-Man	1978–1979	5
Spider-Man	Spider-Man 2	2004	4, 5
Star Trek	Star Trek VI: The Undiscovered Country	1991	2
Star Trek	Star Trek: Nemesis	2003	11
Star Wars			2, 4
Star Wars	Star Wars: A New Hope	1977	9, 11
Star Wars	Star Wars: The Empire Strikes Back	1980	2, 11
Star Wars	Star Wars: Return of the Jedi	1983	11
	Sunshine	2007	8
Superman	Man of Steel	2013	2
	Tarantula	1955	2, 6

Franchise or film series	Film title	Year(s)	Chapter(s)
	Titanic	1997	3
Terminator	Terminator	1984	3
Terminator	Terminator 2: Judgement Day	1991	3
	The Thing	1982	2
	Them!	1954	6
Thor			5
Thor	Thor	2011	3, 9
	Timeline	2003	10
	The Toxic Avenger	1984	3
	Total Recall	1990	6
	Total Recall	2012	4, 5
	Transcendence	2014	7
Transformers			2, 3
	Tron: Legacy	2010	2
	Twister	1996	3
	2001: A Space Odyssey	1969	1, 2, 3, 9
	2010: The Year We Make Contact	1984	4, 9, 11
	Weird Science	1985	2
	What Dreams May Come	1998	10
	What the #$*! Do We Know!?	2004	7
	When Worlds Collide	1951	11
X-Men			2, 3, 6, 7
X-Men	X-Men: First Class	2011	9
	Yesterday Was a Lie	2008	7

Appendix B. TV/Web Series References

Franchise or film series	Series/film title	Episode	Year(s)	Chapter(s)
	Alien Nation		1989–1990	2
	Alphas		2011–2012	3
	Andromeda		2000–2005	9
	Ascension		2014–	9
Babylon 5	Babylon 5		1993–1998	2, 9, 11
Babylon 5	Babylon 5	The River of Souls	1998	3
Babylon 5	Babylon 5	Passing through Gethsemene	1995	8

Franchise or film series	Series/film title	Episode	Year(s)	Chapter(s)
Babylon 5	Crusade		1999	9
Battlestar Galactica	Battlestar Ga-lactica		1978	2, 3, 4, 5, 6, 8, 11
Battlestar Galactica	Battlestar Ga-lactica	The Living Legend	1978	9
Battlestar Galactica	Battlestar Ga-lactica		2004–2009	1, 2, 3, 4, 5, 8, 9, 11
Battlestar Galactica	Battlestar Ga-lactica	Scar		2, 11
Battlestar Galactica	Battlestar Ga-lactica	Valley of Darkness	2005	3
Battlestar Galactica	Battlestar Ga-lactica	No Exit	2009	6
Battlestar Galactica	Battlestar Ga-lactica	Rapture	2007	6
Battlestar Galactica	Battlestar Ga-lactica	Maelstrom	2007	6
Battlestar Galactica	Battlestar Ga-lactica	The Hand of God	2005	11
Battlestar Galactica	Caprica		2009–2010	2, 3, 11
Battlestar Galactica	Battlestar Galac-tica: Blood and Chrome		2012	3
	Blake's 7		1978–1981	1, 7, 11
	The Big Bang Theory		2007–	2, 3, 4, 6
	The Big Bang Theory	The Codpiece Topology	2008	3
	Breaking Bad		2008–2013	2
	City Hospital		1951–1953	2
	Chuck		2007–2012	3
	Continuum		2012–	9, 10
	Cosmos		1980	4, 8, 9
	Cosmos: A SpaceTime Odys-sey		2014	2
CSI				2
CSI	CSI:Crime Scene Investigation		2000–	2, 3
CSI	CSI: Crime Scene Investigation	Overload	2001	4
CSI	CSI: Miami		2002–2012	3

Franchise or film series	Series/film title	Episode	Year(s)	Chapter(s)
	Dallas		1978–1991; 2012–2014	10
	The Day After		1983	6
	Defiance		2013	2, 3, 7, 8, 9, 11
Doctor Who	Doctor Who		1963–1989; 2005–	1, 2, 3, 7, 9, 10, 11
Doctor Who	Doctor Who	The Caves of Androzani	1984	2
Doctor Who	Doctor Who	Terror of the Autons	1971	3
Doctor Who	Doctor Who	Genesis of the Daleks	1974	3
Doctor Who	Doctor Who	Four to Doomsday	1982	4
Doctor Who	Doctor Who	Twin Dilemma	1984	6
Doctor Who	Doctor Who	Into the Dalek	2014	7
Doctor Who	Doctor Who	The Impossible Planet	2006	7
Doctor Who	Doctor Who	The Satan Pit	2006	7
Doctor Who	Doctor Who	The End of the World	2005	8
Doctor Who	Doctor Who	The Pandorica Opens	2010	10
Doctor Who	Doctor Who	Army of Ghosts	2006	10
Doctor Who	Doctor Who	A Good Man Goes to War	2011	11
	Downton Abbey		2010–	2
	Dr. Horrible's Sing-Along Blog		2008	3
	Earth: Final Conflict		1997–2002	11
	Earth 2		1994–1995	11
	Eleventh Hour		2008	2
	ER		1994–2009	2, 3
	Eureka		2006–2012	2, 3, 4, 9, 10, 11
	Eureka	Bad to the Drone	2008	3
	Eureka	Once in a Lifetime	2006	4, 10
	Eureka	Sight Unseen	2007	6
	Eureka	I Do Over	2008	6
	Eureka	Reprise	2011	7

Franchise or film series	Series/film title	Episode	Year(s)	Chapter(s)
	Eureka	O Little Town	2010	7
	Eureka	Force Quit	2012	10
	Eureka	Insane in the p-Brane	2009	10
	Falling Skies		2011–2015	2, 3, 4
	Farscape		1999–2003;2004	2, 9, 11
	FlashForward		2007	2, 3, 4
	Firefly		2002	2, 9, 11
	Freaks and Geeks		1999–2000	3
	Fringe		2008-2013	2, 3, 9, 10
	Fringe	Olivia. In the Lab. With the Revolver	2010	7
	Game of Thrones		2011–	2
	General Hospital		1963–	2
Halo				3
Halo	Halo	Forward Unto Dawn	2012	20
	Hannibal		2013–	3
	Hannibal	Mukozuka	2014	3
	Happy Days		1974–1984	3
	Helix		2014–	2,3
	Hill Street Blues		1981–1987	2,3
The Hitch-Hiker's Guide to the Galaxy	The Hitch-Hiker's Guide to the Galaxy		1981	1, 2, 4, 8, 9
	House of Cards		2013	2
	House, M.D.		2004–2012	2, 3
	The Immortal		1969	2
	The Jimmy Stewart Show		1971	3
	L.A. Law		1986–1994	3
	Logan's Run		1977–1978	3
	Lost		2004–2010	2
	Lost in Space		1965–1968	20
	Marcus Welby, M.D.		1969–1976	2
NCIS				2
NCIS	NCIS		2003–	2, 3

Franchise or film series	Series/film title	Episode	Year(s)	Chapter(s)
	Other Space		2015–	10
	Orange is the New Black		2013–	6
	Pinky and the Brain		1995–1998	3
	Pinky and the Brain	Of Mouse and Men	1995	7
	Primeval		2007–2013	10
	Quark		1977	4
	Red Dwarf		1988–	5, 7
	Red Dwarf	The Last Day	1989	4
	seaQuest DSV		1993–1996	2, 11
Sharknado			2013–	2
Sharknado	Sharknado 2: The Second One		2014	7
	The Shield		2002–2008	2
	Sliders		1995–2000	9, 10
	Space: 1999			1, 9, 10
	Space: 1999	Breakaway	1975	5
	Space: 1999	Death's Other Dominion	1975	11
Stargate				9, 11
Stargate	Stargate SG-1		1997–2007	2, 3, 7, 9
Stargate	Stargate SG-1	Meridian	2002	6
Stargate	Stargate SG-1	A Matter of Time	1999	9
Stargate	Stargate SG-1	Point of View	1999	10
Stargate	Stargate SG-1	Demons	1999	11
Stargate	Stargate Infinity		2002–2003	9
Stargate	Stargate Atlantis		2005–2009	2, 7, 8, 9
Stargate	Stargate Universe		2009–2011	9
	The Starlost		1973–1974	9
Star Trek				1, 2, 3, 7, 9
Star Trek	Star Trek		1966–1969	2, 3, 9, 11
Star Trek	Star Trek	The Ultimate Computer	1968	3
Star Trek	Star Trek	The City on the Edge of Forever	1967	4
Star Trek	Star Trek	The Doomsday Machine	1967	4

Franchise or film series	Series/film title	Episode	Year(s)	Chapter(s)
Star Trek	Star Trek	Errand of Mercy	1967	5
Star Trek	Star Trek	Miri	1966	10
Star Trek	Star Trek	Mirror, Mirror	1967	10
Star Trek	Star Trek	This Side of Paradise	1967	11
Star Trek	Star Trek: The Animated Series	The Terratin Indicent	1973	7
Star Trek	Star Trek: The Next Generation		1987–1994	3, 4, 5, 9, 10
Star Trek	Star Trek: The Next Generation	The Royale	1988	5
Star Trek	Star Trek: The Next Generation	The Enemy	1989	6
Star Trek	Star Trek: The Next Generation	Who Watches the Watchers?	1989	7
Star Trek	Star Trek: The Next Generation	First Contact	1991	7
Star Trek	Star Trek: The Next Generation	Second Chances	1993	7
Star Trek	Star Trek: The Next Generation	Relics	1992	7, 11
Star Trek	Star Trek: Deep Space Nine		1993–1999	4, 9, 10
Star Trek	Star Trek: Deep Space Nine	Our Man Bashir	1995	7
Star Trek	Star Trek: Deep Space Nine	The Ascent	1996	11
Star Trek	Star Trek: Deep Space Nine	In Purgatory's Shadow	1997	11
Star Trek	Star Trek: Voyager		1995–2001	3, 4, 7, 8, 9
Star Trek	Star Trek: Voyager	Parallax	1995	7
Star Trek	Star Trek: Enterprise		2001–2005	2, 10
	Survivor		2000–	5
	Temple Grandin		2010	3
Terminator	Terminator: The Sarah Conner Chronicles		2008–2009	3, 9
Terminator	Terminator: The Sarah Conner Chronicles	Today is the Day pt I	2009	9

Franchise or film series	Series/film title	Episode	Year(s)	Chapter(s)
Terminator	Terminator: The Sarah Conner Chronicles	Today is the Day pt II	2009	9
	3rd Rock from the Sun		1996–2001	2
	Threads		1984	6
	Threshold		2005	3
	Thunderbirds are Go			10
	The Time Tunnel		1966–1967	9
	24		2001–	2
	The Twilight Zone		1959–1964	2, 10
	UFO		1969	10
	Virtuality		2009	11
	The Walking Dead		2010–	2
	The X Files		1993–2002	2, 11
	The X Files	War of the Copro-phages	1996	3
	The Zula Patrol		2005–2010	2

Further Reading

Donna Nelson, Kevin Grazier, Jaime Paglia and Sidney Perkowitz, eds.: *Hollywood Chemistry: When Science Met Entertainment* (American Chemical Society, Washington D.C. 2014)

An anthology divided into two sections, Science Reaches the Screen includes chapters describing how accurate science makes its way onto both big and little screen, Science Reaches the Public looks at the effect of Hollywood science on viewers, education, even public policy.

John Baxter: *Science Fiction in the Cinema* (Paperback Library, New York 1970)

An analysis of screen science fiction's origin and evolution, and how and why it is different than literary science fiction.

Christopher Frayling: *Mad, Bad, and Dangerous?: The Scientist in the Cinema* (Reaktion Books, London 2005)

A critical examination of the portrayal of scientists through cinematic history.

David A. Kirby: *Lab Coats in Hollywood* (MIT Press, Cambridge, Massachusetts 2011)

A history of how Hollywood has used scientists as advisors in both TV and film.

Steven Weinburg: *The First Three Minutes* (Basic Books, New York 1977, 1988)

An excellent introduction to our understanding of the hectic moments after the creation of the universe, where cosmology and particle physics merge.

Sean Carroll: *From Eternity to Here: The Quest for the Ultimate Theory of Time* (Penguin, New York 2010).

A wonderful examination of all the physics relevant to an understanding of time (as much as anyone understands it). Very accessible to the lay person.

Gary Zukav: *The Dancing Wu Li Masters: An Overview of the New Physics* (Morrow, New York 1979)

As mentioned in the text, this is a good non-mathematical introduction to quantum physics but beware that the line between the metaphor of eastern mysticism and actual physics sometimes gets blurred.

Paul Gilster: *Centauri Dreams: Imaging and Planning Interstellar Exploration* (Copernicus, New York 2004)

A look at how we might pull off interstellar travel without such time-saving devices as wormholes or warp drives.

..ard Gott: *Time Travel in Einstein's Universe: The Physical Possibilities of Travel Through Time* (Mariner Books, Boston 2002)

A non-mathematical introduction to how Relativity Theory might permit time travel, and presents several different possible types of time machine.

Dennis Overbye: *Lonely Hearts of the Cosmos* (Back Bay Books, Boston 1991)

The story of how we figured out how big and how old the Universe is.

Jeremy Bernstein: *Nuclear Weapons: What You Need to Know* (Cambridge University Press, Cambridge 2008)

A slim, but informative, primer on the science, history, and effects of nuclear weapons.

Amanda Gefter: *Trespassing on Einstein's Lawn: A Father, a Daughter, the Meaning of Nothing, and the Beginning of Everything* (Bantam Books, New York 2014)

A non-physicist sets out to understand how something could have come from nothing, and gets what answers she can from some of the best theoretical physicists currently around.

Philip Plait: *Death From the Skies!: The Science Behind the End of the World* (Viking, New York 2008)

The wonder and majesty of our cosmic neighborhood—and all the ways it might kill us.

CPSIA information can be obtained
at www.ICGtesting.com
Printed in the USA
LVOW06s0332200717

541902LV00003B/14/P